RECENT ADVANCES AND ISSUES IN
Meteorology

**Recent Titles in the
Oryx Frontiers of Science Series**

Recent Advances and Issues in Chemistry
David E. Newton

Recent Advances and Issues in Physics
David E. Newton

Recent Advances and Issues in Environmental Science
Joan R. Callahan

Recent Advances and Issues in Biology
Leslie A. Mertz

Recent Advances and Issues in Computers
Martin K. Gay

Oryx Frontiers of Science Series

RECENT ADVANCES AND ISSUES IN Meteorology

Amy J. Stevermer

ORYX PRESS
Westport, Connecticut • London

The rare Arabian Oryx is believed to have inspired the myth of the unicorn. This desert antelope became virtually extinct in the early 1960s. At that time, several groups of international conservationists arranged to have nine animals sent to the Phoenix Zoo to be the nucleus of a captive breeding herd. Today, the Oryx population is over 1,000, and over 500 have been returned to the Middle East.

Library of Congress Cataloging-in-Publication Data

Stevermer, Amy J.
    Recent advances and issues in meteorology / Amy J. Stevermer.
      p. cm.—(Oryx frontiers of science series)
    Includes bibliographical references and index.
    ISBN 1–57356–301–3 (alk. paper)
    1. Meteorology. I. Title. II. Series.
QC861.3.S74 2002
551.5—dc21   2001034826

British Library Cataloguing in Publication Data is available.

Copyright © 2002 by Amy J. Stevermer

All rights reserved. No portion of this book may be reproduced, by any process or technique, without the express written consent of the publisher.

Library of Congress Catalog Card Number: 2001034826
ISBN: 1–57356–301–3

First published in 2002

Oryx Press, 88 Post Road West, Westport, CT 06881
An imprint of Greenwood Publishing Group, Inc.
www.oryxpress.com

Printed in the United States of America

The paper used in this book complies with the Permanent Paper Standard issued by the National Information Standards Organization (Z39.48–1984).

10 9 8 7 6 5 4 3 2 1

Copyright Acknowledgments

The author and publisher gratefully acknowledge permission for use of the following material:

American Meteorological Society policy statements, "The Public/Private Partnership in the Provision of Weather and Climate Services" (July 23, 1999) and "Planned and Inadvertent Weather Modification" (October 2, 1998). Used by permission.

Roger A. Pielke, Jr., "Six Heretical Notions about Weather Policy," *Weatherzine* (April 2000). Used by permission.

Excerpts from Timothy E. Wirth speech delivered to the National Academy of Sciences on November 17, 1997. Courtesy of U.S. Department of State.

Excerpts from World Meteorological Society, "Scientific Assessment of Ozone Depletion: From the 1998 Assessment Executive Summary." Courtesy of the WMO.

Excerpts from Robert T. Watson speech delivered to the United Nations Framework Convention on Climate Change on November 2, 1999. Courtesy of the Intergovernmental Panel on Climate Change.

Excerpts from Intergovernmental Panel on Climate Change, "Second Assessment Synthesis of Scientific-Technical Information." Courtesy of the IPCC.

Contents

Preface   ix
Acknowledgments   xiii

Chapter 1. Meteorology Today   1
Chapter 2. Meteorological Technology   61
Chapter 3. Meteorology and Society   93
Chapter 4. Unsolved Problems, Unanswered Questions   125
Chapter 5. Biographical Sketches   141
Chapter 6. Career Information   155
Chapter 7. Statistics and Data   189
Chapter 8. Selected Reprints, Documents, and Reports   197
Chapter 9. Professional Societies and Research Organizations   235
Chapter 10. Print and Electronic Resources   245
Chapter 11. Glossary   261

Index   269

Preface

The study of meteorology and atmospheric science encompasses many large and sometimes controversial issues, including long- and short-term forecasting, climate prediction, climate change, ozone depletion, and air pollution. This book focuses on the latest research and developments in these areas and many more. Every attempt has been made to convey the voice of the meteorological community as a whole, rather than any one personal opinion. The material presented here was drawn from professional scientific journals, reports, newspaper articles, textbooks, and other written and electronic media, most of it peer reviewed and much of it in the public domain. Complete bibliographic information accompanies each section of the book, and a comprehensive listing of print and nonprint resources is provided in Chapter 10.

The book examines issues in the various subdisciplines of meteorology and the atmospheric sciences. Each chapter is organized into main topic sections around these subdisciplines, and the main topic sections are further broken down into subsections. For instance, Chapter 1 contains a main topic section "Atmospheric Chemistry," which includes subsections on stratospheric ozone and tropospheric chemistry issues. The latter is further divided into sections on acid rain and on visibility and air quality.

Scientific units remained a potentially unending challenge throughout the writing and editing process. The convention used throughout the book is to convey units in metric, with the English equivalent in parentheses. This method was selected to convey units of measure as clearly as possible.

This book will provide readers, especially nonmeteorologists, with an overview of the latest advances in the meteorological sciences. Focus is on research and developments of the late 1990s, with particular emphasis on topics of large national and global interest. This book provides a look at the challenges facing scientists, policymakers, and the public in the twenty-first century. As we face possible climate changes induced by our own human actions, the atmosphere and environment are sure to remain topics of much concern in the years to come.

Chapter 1 introduces the subject of meteorology and the atmospheric sciences and provides an overview of some of the recent research in the discipline. The chapter covers three broad areas: weather forecasting, atmospheric chemistry, and climate change. Topics addressed include the status of numerical weather prediction, current events and knowledge concerning severe storms and other phenomena, progress in understanding El Niño and La Niña, issues in air quality, and greenhouse warming and climate change.

Chapter 2 presents many of the technologies used to study the atmosphere, its components, and movements. The material addresses the role of computers and satellites and profiles new instruments and techniques used to measure atmospheric quantities and behavior. Several field experiments involving suites of instruments and researchers from countries worldwide are also discussed.

Chapter 3 seeks to illuminate societal impacts of, and on, weather and climate. Some of the material focuses on hazards: the personal and economic costs of severe weather or climate extremes, for example. Other sections examine the debate between private sector and government forecasting, predictions of climate change and current issues regarding mitigation and adaptation, and weather modification.

Chapter 4 addresses unanswered questions and examines the future of meteorology. The chapter presents some of the community's recommendations for future research and direction in the atmospheric sciences. These issues are discussed according to their subdiscipline within meteorology, including atmospheric dynamics, atmospheric chemistry, and climate.

Chapter 5 offers biographies of many of the scientists referenced in this book. The biographies clearly show that people working in meteorology and the atmospheric sciences come from many different backgrounds and focus their energies in many areas.

Chapter 6 is for persons interested in finding out more about a meteorological career. The chapter discusses the education and background typically required for most positions in meteorology, from operational meteorology to television weathercasting to careers in research. The chapter also profiles three people working in the profession and suggests resources for additional information.

Chapter 7 presents statistics related to various aspects of meteorology, including information on salaries and employment in meteorology, data on atmospheric sciences funding, and some statistics on severe weather.

Chapter 8 contains reprints from assessment reports, policy statements, and other documents related to key weather and climate issues. Chapter 9 lists various meteorological societies and professional organizations, as well as current private sector weather vendors in the United States. Chapter 10 contains lists of print and nonprint resources that provide additional information on atmospheric topics. Chapter 11 contains a glossary of terms.

Given the pace of technological developments, it should be noted that some of the specifics mentioned in relation to forecast models, satellite instrumentation, and other areas may be out of date by the time this book is published. Model resolutions can change, and budget constraints can determine which satellites fly and which do not, to name just a couple of examples. At the time this goes to press, every effort has been made to ensure that the material presented is scientifically and technically accurate.

Acknowledgments

This book would not have been possible without the comments and contributions of many individuals in the atmospheric sciences community. Tremendous thanks are extended to these people, and particularly to Anthony Rockwood, Ed Szoke, Eric Thaler, Ed Zipser, Walter Robinson, Bob Glancy, Josh Wurman, Nita Fullerton, Betsy Weatherhead, Bryan Johnson, Roger Pielke, Jr., Marty Venticinique, Coleen Decker, Pam Daale, John Toohey Morales, Bill Tahnk, Eugenia Kalnay, Will von Dauster, and the staff at the American Meteorological Society.

I extend special thanks to the people who have guided my own education and development in the atmospheric sciences and science communication fields, and to Henry Rasaf, John Wagner and the production staff at Oryx Press and Greenwood Publishing Group for their work finalizing the manuscript for publication.

Chapter One

Meteorology Today

> We live in an atmosphere that shapes our activities and states of mind, an atmosphere whose storms threaten our lives and property, and whose climate and composition influence the nature and vitality of our societies.
> —National Research Council, 1998

The Greek term *meteor* is used to refer to anything in the air. *Meteorology* is defined as the study of the phenomena of the atmosphere, including atmospheric motions, atmospheric physics, and atmospheric chemistry. It also includes the atmosphere's interactions with and effects on land surfaces, oceans, and life in general. This study of the atmosphere is usually closely combined with other branches of physical science, so that the subjects of meteorology may also be described using the more comprehensive term *atmospheric sciences.*

Meteorology and the atmospheric sciences have truly come of age in recent decades, with significant advances in weather forecasting, climate research, and our understanding of turbulence and air pollution. Although we may not often think about it, the thin blanket of air that is earth's atmosphere affects every being on the planet. It is only natural, therefore, that so much expertise and effort should be devoted to understanding the dynamics, chemistry, heat exchange, and other processes of this body of air. In this chapter, we look at recent advances in some of the major topics in meteorological and atmospheric research. These topics have been divided into three broad categories related to weather forecasting, atmospheric chemistry, and climate research.

THE ATMOSPHERE

The earth's atmosphere is a thin layer of trillions and trillions of gas molecules, of which approximately 78 percent are nitrogen and 21 percent are oxygen. Argon, neon, helium, hydrogen, and xenon are present in trace amounts, and carbon dioxide, methane, nitrous oxide, ozone, and water vapor are present in variable amounts. Although this layer extends upward for hundreds of kilometers, 99 percent of it lies within 30 kilometers (20 miles) of the earth's surface.

The air molecules taken in total weigh approximately 5600 trillion tons. This weight exerts a force on the earth, referred to as *pressure*. The pressure at any level in the atmosphere relates to the total weight of the air above that point; atmospheric pressure therefore decreases with altitude as the number of molecules above a particular point becomes fewer. The average surface pressure is 1013.25 millibars (mb), also referred to as 1 atmosphere (atm). The actual pressure at a particular location depends on the elevation of the location and on the physics and dynamics of the atmosphere. Horizontal variations in atmospheric pressure are extremely important, as it is these pressure differences that give rise to winds.

Air temperature also varies with height in the atmosphere, and these temperature variations are the basis for the different atmospheric layers shown in Figure 1.1. The layer closest to the surface of the earth, to about 11 kilometers (7 miles) up, is referred to as the *troposphere*. All of the weather we experience occurs in this layer, where air temperature normally decreases with height as the sun warms the earth's surface. Rising, warmer air and descending, cooler air keep this layer very well mixed: its very name, from the Greek *tropein*, means to turn or change.

The boundary between the troposphere and the layer above it is called the *tropopause*. The altitude of the tropopause, marked by the level at which temperature stops decreasing with height, varies from equator to poles and with time of year. The tropopause boundary generally keeps the air below it from mixing with the air above it. Exceptions are breaks in the tropopause, which can mark the position of high winds rushing along at speeds exceeding 190 kilometers (115 miles) per hour. These winds are known as jet streams.

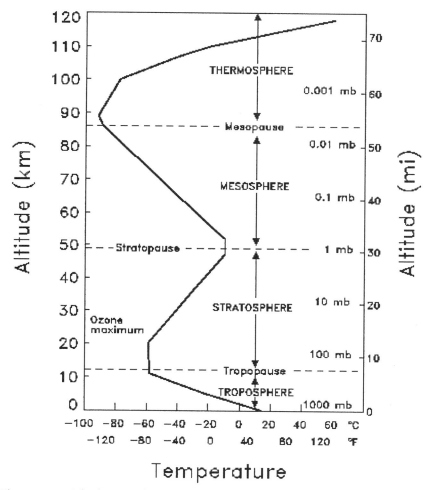

Figure 1.1. The layers of the atmosphere are defined in terms of the vertical profile of temperature. Weather as we know it occurs in the lowermost layer, the troposphere. The layer above the troposphere, known as the stratosphere, is home to the protective ozone layer. Together, these two layers comprise 99.5 percent of the mass of the earth's total atmosphere. Figure provided by W.R. Tahnk, Oregon State University.

Above the tropopause is the *stratosphere*, or stratified layer. Here, air temperature increases with height, suppressing vertical mixing. The stratosphere contains the ever-essential ozone layer, located between 20 and 30 kilometers (12 and 19 miles) above the earth's surface. Ozone absorbs large amounts of ultraviolet (UV) radiation from the sun. Some of the absorbed energy warms the stratosphere, explaining the resulting temperature profile.

At about 50 kilometers (30 miles) above the surface is the *stratopause*, which separates the stratosphere from the *mesosphere*. At the altitude of the mesosphere, air pressure is quite low, and air temperatures are the coldest of the entire atmosphere, averaging −90°C (−120°F). Above the mesosphere is the *thermosphere*. In this layer, solar energy absorbed by oxygen molecules warms the air. Air molecules are so sparse here that a single molecule would have to travel an average distance of over half a mile to collide with its closest neighbor. At the top of the thermosphere, this distance is even greater. Above the thermosphere, the gravitational force is so weak that molecules can actually escape to space. This region, sometimes referred to as the *exosphere*, represents the upper limit of the atmosphere.

Some scientists concentrate their concern on an additional region of the atmosphere, extending from about 60 kilometers (40 miles) altitude on up to the exosphere. This region, called the *ionosphere*, contains fairly large concentrations of positively charged atoms, called ions, and free electrons. By reflecting standard AM radio waves back to earth, this region plays a major role in communications.

This introduction to the structure and regions of the atmosphere should provide some sense of the vastness and complexity of its science. It will also help provide perspective for many of the discussions that follow, including stratospheric ozone depletion, weather forecasting, and global warming. Historically, meteorologists have been concerned primarily with the troposphere, and in recent decades, the stratosphere. The material in this book follows current advances in the science of these two layers, which together contain almost 100 percent of the total atmosphere.

WEATHER FORECASTING AND FORECAST MODELS

Although most of us are familiar with the daily weather forecasts obtained from television, radio, newspapers, or the Internet, the science behind these views of the future may be something of a mystery. While meteorologists may know lots about the science going into the forecasts, being able to predict future weather accurately is often a mystery to them as well. Meteorologists do not have crystal balls to provide reasonable weather forecasts; instead they rely on being able to simulate atmospheric behavior accurately in computer models, called *numerical weather prediction* models. Quality observations of current weather conditions are required to provide correct initial conditions for these models.

In essence, forecasting the weather requires knowledge of what is upstream from you and some understanding of how conditions may evolve as they make their way toward you. Neither of these is easy. With much of the earth's surface covered by ocean, where observations are limited, timely and accurate weather information for many locations is not a simple possibility. This lack of data makes a big difference when it comes to producing a reliable forecast. A 24-hour forecast for Boston, for example, requires knowledge of conditions across all of North America. A forecast for one week ahead requires data from locations across the entire globe. New technologies have done much to improve the availability of worldwide data, and we discuss them in Chapter 2. Still, the observations are only part of the puzzle. TV meteorologist Paul Douglas relates weather forecasting to attempting to predict exactly where a twig will be five days after it is dropped into a stream. Just as there is a nearly endless list of conditions and events influencing the final destination of the twig, large numbers of factors affect the movement of storms and other weather variables. It is precisely these factors with which weather forecast models are concerned.

In the 1870s, the Army Signal Corps began production of the first regular daily weather maps and forecasts in the United States. The forecasts were made by plotting symbols on maps to indicate meteorological conditions such as temperature and pressure. Experienced meteorologists could then use this information to attempt to predict how conditions would evolve. Since then, the science and technology of weather forecasting have changed significantly. The 1940s and 1950s saw the establishment of an almost global balloon-borne instrument, or radiosonde, network. This network greatly increased the amount of information available to meteorologists by adding important observations of wind, temperature, and moisture above the earth's surface. In the late 1950s, advances in computer technology led to the beginnings of numerical weather prediction, though at a far more primitive level than is associated with numerical weather prediction today. The 1950s also brought radar data to the weather arena, and with the launch of weather satellites beginning in the 1960s, improved accuracy in weather forecasting seemed a tangible possibility.

Meteorology is a science in which sharing data and research improvements has been a long tradition. This tradition has allowed progress to occur on many fronts, and many countries to benefit from this progress. Weather forecasting, with its dependence on global observational data, relies heavily on such collaborations. To be anything approaching successful, modern forecasting requires access to both observations and a shared knowledge of how to solve the complicated equations of motion of the atmosphere. The tools used for this feat are weather forecast models, and their development and improvement has been pursued internationally.

Weather forecast models are computer projections of the conditions of the atmosphere at a given time in the future. These models use mathematical equations to predict variables such as temperature, moisture, and wind at several heights in the troposphere. Using input data from weather balloons, surface observations, aircraft, and satellites, the models simulate the state of the atmosphere anywhere from a few hours to several months in the future.

The first operational weather forecast model to be implemented in the United States was the *barotropic* model. In a barotropic atmosphere, levels of equal atmospheric density are considered to be parallel to levels of equal atmospheric pressure. Introduced in 1958, the barotropic model was a reduction from a three-level model developed three years earlier by Jule Charney at the Institute for Advanced Studies in Princeton. Because computers at that time lacked the necessary processing speed and memory, the barotropic model did not have the vertical or horizontal resolution to account for the transport, or *advection*, of temperature. Instead of associating the pressure at a certain altitude with a temperature value, the heights of pressure levels depended on the amount of spin, or vorticity, associated with a column of air. In meteorology, vorticity usually refers to the amount of spin around a vertical axis, and, to simplify substantially, has opposite directions for high and low pressure. By associating vorticity with pressure, the barotropic model provided a fairly adequate representation of large-scale atmospheric processes. Nevertheless, the bulk of forecast generation was still done manually by meteorologists who plotted maps and visually examined the location of high and low pressure areas and other features. It was not until the 1970s that the National Weather Service adopted the barotropic model for real-time operational use, thus ensuring a major role for numerical weather prediction in future forecasting methods.

In 1971, the Limited-area Fine Mesh (LFM) model was implemented as the first regional model used by the National Meteorological Center (now NCEP). The LFM, a significant advancement from the relatively simple barotropic model, was able to solve the full equations of motion, largely due to the increase in computer power available by that time. The model remained in use for over twenty years and was the basis for Model Output Statistics (MOS), a set of statistical equations used to generate probabilities for forecast variables such as precipitation and expected high and low temperatures. In the late 1970s, a new model, the nested grid model (NGM), was introduced as a replacement for the barotropic model. This new model consisted of three nested grids (hence the name), the largest of which was a coarse grid covering the entire globe. Within this grid was a subgrid over a hemisphere-sized area, and within the subgrid, an even finer grid over an area about the size of North America. The improved coverage and resolution of this model offered many advantages. For the first time ever, model calculations could be performed using knowledge of what was upstream. Since weather affecting the western United States moves ashore from the Pacific basin, this type of information was especially essential. The NGM is still being run today, having been continually improved and modified through the mid-1980s, when focus was shifted to development of newer models.

The true explosion in numerical weather prediction occurred during the 1990s and is attributable largely to the increase in computing capabilities accompanied by new developments in numerical models and their initialization. With today's high-speed computer network connections, as well as the ability to transfer data by satellite via the Satellite Broadcast Network (SBN), forecasters can view results from a number of models with various spatial and temporal resolutions. Figure 1.2 shows the time coverage associated with many of the major forecast models in use. And because all model output is part of the public domain, anyone can view and attempt to interpret the model results. The ability to access this information has meant big changes for newcomers to the field. Meteorologists being trained today must have a solid understanding of the physics of the models as well as the models' limitations. Increasingly, this knowledge will be applied to interpreting the models and doing what the models cannot—communicating information to the public, especially when the information pertains to weather with the potential of endangering lives.

The weather forecast models used today are divided into three classes: global models, regional models, and local-scale models. Global models predict events over the Northern Hemisphere or entire globe. Global models include the medium-range forecast model (MRF), the U.S. Navy's NOGAPS model, and the Global Ocean Model. In Eu-

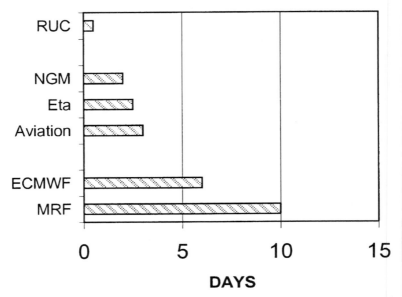

Figure 1.2. Time coverage of forecast models currently in use. The Rapid Update Cycle (RUC) model is a local-scale model. The nested grid model (NGM), Eta, and Aviation models provide regional coverage, and the medium-range forecast model (MRF) and European Centre for Medium-Range Weather Forecasts' model (ECMWF) are global models. All the models require large computing resources to run.

rope, global models include those run by the European Centre for Medium-Range Weather Forecasts and by the United Kingdom Meteorological Office. Canada has its own global model, known as the Global Spectral Model. These models require extremely powerful supercomputers and are run at only a few sites in the United States and internationally.

The next group of models focuses on an area about the size of the United States. These are the regional models and in the United States are operated mainly by the National Centers for Environmental Prediction (NCEP). Forecast models provide predictions of weather conditions at a predefined spatial resolution. Spatial *resolution* corresponds to the size of the area for which observable quantities can be separated and distinguished. In general, the smaller the model resolution, the longer the model will take to run on the computer or the more powerful the computer must be to run the model in a timely manner. It is not likely, for instance, to obtain forecast information specifically for your own yard. The NGM is a regional model providing a spatial resolution of 80 kilometers (50 miles). That is, the predictions obtained by this model are valid for a grid of squares each measuring 80 kilometers by 80 kilometers—definitely larger than most people's yards. The NGM is run twice daily and provides forecasts out to 48 hours at 6-hour intervals.

Other regional models currently in use include the Eta model and the aviation model. The Eta model is named after a mathematical coordinate system able to account more accurately for mountain ranges, river valleys, and other topography. Eta has a 32-kilometer (20-mile) spatial resolution and also has its own Model Output Statistics. According to former NCEP director Ronald McPherson, Eta has outperformed everything in its ability to predict precipitation amounts. Eta is run four times daily, providing forecasts out to 36 and 48 hours for time intervals as short as hourly. The model output is available at finer resolution than most forecasters see, due to current limitations in what can be transmitted via high-speed links or by satellite. Another regional model, the aviation (AVN) model, is the short-range part of the Global Spectral Model. It is run twice daily and provides forecasts out to 72 hours at 6-hour time intervals. With a spatial resolution of 100 kilometers (60 miles) and its own set of Model Output Statistics, the aviation model was developed for use primarily by the aviation industry.

Forecast models have been continually updated and improved. As both computer power and meteorologists' level of understanding increase, there has been a large push to increase the accuracy of long-range forecasts. These forecasts currently provide reasonably accurate information out to three or five days in the future, but a more accurate seven-day forecast may be coming soon. In 1992, a new technique to improve long-range forecast accuracy was initiated at the NCEP in the United States and at the European Centre for Medium-Range Weather Forecasting. The technique, called *ensemble modeling*, uses multiple model runs with perturbations introduced at key areas. Because a small change early on can lead to dramatic differences at longer timescales, ensemble modeling provides a set of results that can be studied to determine the most stable solution. Because the components of the forecast that are most uncertain tend be averaged out, ensemble modeling produces a forecast that for time periods beyond a few days can be much more accurate than individual forecasts. One of the newest trends in ensemble modeling is to make an ensemble of a small number of ensembles from the different models—

for example, the medium-range forecast (MRF) model and results from the European Centre for Medium-Range Weather Forecasts' model. The technique allows small changes within the models to be minimized, so that an average, or most likely, view of future conditions can be obtained.

Other emphasis has been on the development of local-scale models to provide detailed analyses and forecasts of meteorological phenomena ranging from a few kilometers to about 100 kilometers (60 miles) in size. This scale, referred to as the *mesoscale*, includes thunderstorms and other smaller-scale systems like sea breezes. These models can account for terrain and other fine-scale features and are often tailored to specific regions. An example of a mesoscale model is the MM5, a community mesoscale model in its fifth generation of development. Originated by the Pennsylvania State University and the National Center for Atmospheric Research, MM5 is able to provide forecast output for spatial scales of only a few kilometers. At this scale, weather forecasts could be issued that would be specific to particular neighborhoods. The model is being continually improved by contributions from users at several universities and government laboratories.

Recent developments in short-range weather forecasting also include the Rapid Update Cycle (RUC). RUC forecasts are initiated with very recent observational data from commercial aircraft reports and from profilers and other instruments. The aircraft data are provided through the Aircraft Communications Addressing and Reporting System (ACARS), discussed in greater detail in Chapter 2. RUC's use of data collected outside the time of the twice-daily radiosonde releases is its key difference from the other models. The model has its basis on work done by the National Oceanic and Atmospheric Administration's Forecast Systems Laboratory during the 1980s and represents the first national model developed outside the National Centers for Environmental Prediction. The RUC provides forecast guidance in the 0- to 12-hour time frame. RUC output is especially useful for predicting short-term weather events and also has applications for severe weather and for aviation forecasting. The first version was implemented by NCEP in September 1994 and was run every three hours at a 60-kilometer (40-mile) spatial resolution. A new version, RUC-2, became operational in April 1998 and is run every hour at a 40-kilometer (25-mile) resolution. As computers continue to become faster and more powerful, future versions of the RUC and other operational models will be able to provide predictions at increasing smaller spatial scales.

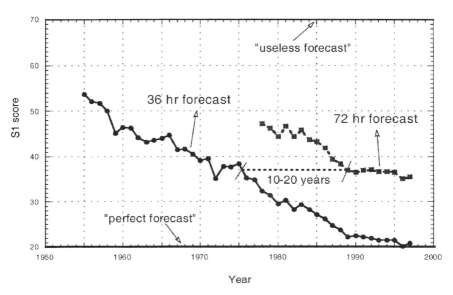

Figure 1.3. Improvement in skill scores for 36-hour and 72-hour weather forecasts. The scores indicate that by 1990, meteorologists were able to predict the weather 72 hours into the future with the same accuracy as was associated with a 36-hour forecast just fifteen years before. From Kalnay, Lord, and McPherson. Used by permission of the American Meteorological Society.

As one would expect, continual improvements in weather forecast models at all time and space scales has led to dramatic improvements in the accuracy of predictions. Forecast accuracy is measured quantitatively as a level of skill. The *forecast skill* is a statistical score associated with how well a forecast is able to predict atmospheric pressure at a given height over the entire computational area of the model. On this scale, a score of 70 would apply to a largely worthless forecast, and a score of 20 would indicate a perfect prediction. Figure 1.3 shows the evolution of skill scores since the beginnings of operational numerical weather prediction in the 1950s. Many of the improvements in skill scores are directly attributable to advances in computer technology, as will be discussed in the next chapter. Thirty-six-hour forecasts today are now generally "perfect," at least in the general features, and the 72-hour forecasts of today are as accurate as 36-hour forecasts were just ten to twenty years ago. Five-day forecasts had virtually no useful skill fifteen years ago but are now moderately skillful. Such improvements are attributable to advances in scientists' ability to represent atmospheric processes within the models and to increased computing power, as well as to large increases in data availability for accurate model initializations.

Observations used to initialize weather forecast models come from a variety of sources. With the recent National Weather Service modernization, many observations now come from Automated Surface Observing Systems (ASOS). Radiosondes, measuring profiles of temperature, humidity, and wind, are carried aloft by weather balloons launched from National Weather Service offices two times daily. The National Oceanic and Atmospheric Administration provides hourly averaged vertical profiles of wind speed and direction from a network of thirty-two wind profilers located mostly in the central United States. (These and other new instrumentation are described in greater detail in Chapter 2.)

Overall, the advances that have been and continue to be made in forecasting have done much to improve weather prediction. Warning times have been lengthened to give people more time to seek shelter, and long-range forecasts are becoming more accurate. New research in severe storm prediction and understanding is examined more specifically below. Also discussed are El Niño and La Niña and their roles in affecting world weather.

References

Ahrens, C.D. *Essentials of Meteorology. An Invitation to the Atmosphere*. St. Paul, MN: West, 1993.

Douglas, P. *Prairie Skies. The Minnesota Weather Book*. Stillwater, MN: Voyageur Press, 1990.

Fleming, J.R. *Historical Essays on Meteorology 1919–1995*. Boston: American Meteorological Society, 1996.

Fullerton, N., ed. *FSL in Review 1997–98*. Boulder, CO: National Oceanic and Atmospheric Administration Forecast Systems Laboratory, 1998.

Kalnay, E. Chapter One: Overview from *Numerical Weather Prediction*. 1998. <http://weather.ou.edu/~ekalnay/NWPChapter1/NWPChapt1.html>.

Kalnay, E., S.J. Lord, and R.D. McPherson. "Maturity of Operational Numerical Weather Prediction: Medium Range." *Bulletin of the American Meteorological Society*, 79(12): 2753–2769, 1998.

National Centers for Environmental Prediction, <http://www.ncep.noaa.gov>. Includes a brief history of the Hydrometeorological Prediction Center at <http://www.hpc.ncep.noaa.gov/html/historyNMC.html>.

National Oceanic and Atmospheric Administration Forecast Systems Laboratory, <http://www.fsl.noaa.gov/>.

National Research Council. *The Atmospheric Sciences: Entering the Twenty-first Century.* Washington, DC: National Academy Press, 1998.
PSU/NCAR Mesoscale Modeling System, <http://www.mmm.ucar.edu/science/weather.html>.
Rockwood, A. Personal communication, 1999.
Shuman, F.G. "History of Numerical Weather Prediction at the National Meteorological Center." *Weather and Forecasting,* 4(3): 286–296, 1989.
Szoke, E. Personal communication, 1999.
Thaler, E. Personal communication, 1999.

SEVERE WEATHER

One important application of weather forecasting is its ability to warn people of severe weather. Severe weather can include thunderstorms, tornadoes, hurricanes, blizzards, ice storms, floods, droughts, and extreme heat or cold, all of which can damage human health and property. Meteorologists at the National Severe Storms Laboratory and elsewhere may spend their entire careers working to understand and better predict severe weather. An overview of some of the types of severe weather and new advances in our understanding of them is presented below.

Thunderstorms

Ask people for the form of severe weather they have most often experienced firsthand, and they are likely to name the thunderstorm. An estimated 40,000 thunderstorms occur each day throughout the world. This daily value corresponds to a total of over 14 million thunderstorms annually! Thunderstorms are, as their name implies, storms containing lightning and thunder. They are capable of producing heavy rain, large hail, gusty winds, flash floods, and, in some cases, tornadoes. Especially dangerous thunderstorms, producing three-quarter-inch hail and/or wind gusts of 92 kilometers per hour (57 miles per hour) or more, are characterized by the National Weather Service as severe thunderstorms.

Meteorologists have made considerable progress in understanding the conditions favorable for thunderstorm development. These conditions may be more prevalent in certain geographic regions than in others. For instance, the combination of warmth and moisture makes thunderstorm development particularly likely over the equatorial regions. In these areas, thunderstorms occur on about one of every three days. In the United States, thunderstorms are most frequent in the Southeast—particularly in the Gulf Coast states of Louisiana, Alabama, Mississippi, and Florida. Warm, moist air in these regions results in an average of seventy to ninety thunderstorm-days per year. Thunderstorms rarely occur in dry climates. Lack of available atmospheric moisture in subtropical deserts around 30 degrees latitude leads to an extremely small likelihood for thunderstorm formation in these areas.

Much of what meteorologists know about thunderstorms they have known for a while. This includes the cycle with which thunderstorms progress from birth, through maturity, to decay. For air-mass thunderstorms, these processes are especially well understood. Air-mass thunderstorms, also known as ordinary thunderstorms, tend to develop in the summertime in warm, humid air masses. The life cycles of these thunderstorms are more easily understood since they typically occur away from weather fronts and are usually of short duration. Air-mass thunderstorms form when humid air rises, cools, and condenses into a single cumulus cloud or cluster of clouds. Cumulus clouds are most often described as "puffy" clouds, looking like giant, floating pieces of cotton with a flat, white or light gray base. The cloud grows as long as it is supplied with moist air rising upward from the surface. Eventually the cloud droplets become larger and heavier, and begin to fall. As this happens, air surrounding the cloud is drawn in, and because this air is colder and heavier than the moist cloud air, it begins to descend in a downdraft. The presence of this downdraft, in combination with the upward air motion feeding warmer, moister air into the cloud, marks the establishment of a mature thunderstorm. This is the stage during which an air-mass thunderstorm is most intense: heavy rain, and sometimes small hail, may fall from the cloud; the top of the cloud may extend upward to an altitude of over 12 kilometers (40,000 feet) and be pushed by upper-level winds into an anvil shape. During this stage, the updrafts and downdrafts are at their strongest. The mature stage lasts for about 15 to 30 minutes, until the upward air motion that brings moisture to the cloud begins to weaken. The storm is deprived of its moisture, and therefore its source of energy, and eventually dissipates and dies.

The thunderstorms capable of significant damage to life or property are usually classified as severe. As with air-mass thunderstorms, their formation is a fairly well-understood process. In general, severe thunderstorms form similarly to an air-mass thunderstorm, as warm, moist air is forced upward in unstable air to cool and condense into cloud droplets. However, associated with severe storms is an area of strong vertical wind shear, which occurs when the winds increase or change direction with height. Shear can cause the storm system to tilt. This tilting makes it possible for the updrafts to grow separately from the downdrafts, allowing the updrafts to continue to supply moisture to the cloud. The thunderstorm is therefore able to maintain itself for longer time periods and gain enough energy to potentially become violent.

Severe thunderstorms are generally categorized into two types: supercell storms and squall line storms. A *supercell* storm is an enormous thunderstorm whose upward and downward air motions are essentially in balance, allowing the storm to remain as a single, nondecaying entity for a period of hours. A single supercell storm can produce damaging winds, grapefruit-size hail, and large, long-lived tornadoes. In contrast, a *squall line* is a line of several severe thunderstorms, often occurring up to 300 kilometers (over 180 miles) ahead of a cold front. Such a line is formed by complex dynamic conditions in the earth's atmosphere, such as air aloft flowing over the cold front or a collision of distinct air masses. These thunderstorms are especially likely to occur during springtime over the Central Plains of the United States.

Other types of severe weather include tornadoes and hurricanes, as well as flash floods, hail storms, and winter storms and blizzards. In recent years, population growth and increased development have made people more vulnerable to severe weather. These factors have made accurate weather forecasting especially essential, as discussed further in Chapter 3. The following sections examine some of the extreme weather events that have occurred recently and highlight new knowledge that has been attained.

References

Ahrens, C.D. *Essentials of Meteorology: An Invitation to the Atmosphere*. St. Paul, MN: West, 1993.

Lutgens, F.K., and E.J. Tarbuck. *The Atmosphere: An Introduction to Meteorology*. 4th ed. Englewood Cliffs, NJ: Prentice Hall, 1989.

National Severe Storms Laboratory, <http://www.nssl.noaa.gov>.

Tornadoes

Thunderstorms, especially supercell thunderstorms, can result in the formation of a tornado. *Tornadoes* are rapidly rotating winds that blow around a small area of intense low pressure. Most last only a few minutes and have a path length of only a few kilometers. How-

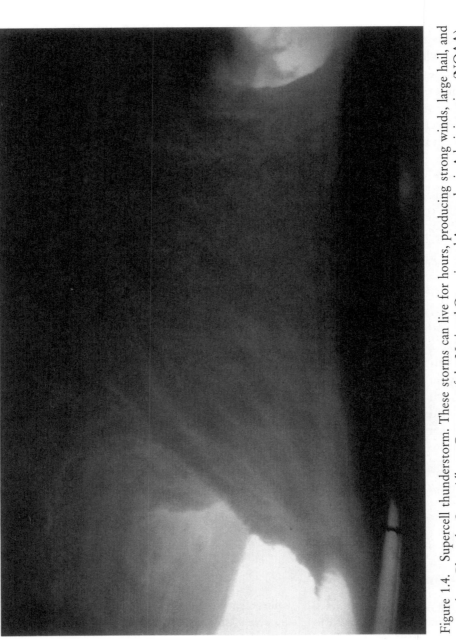

Figure 1.4. Supercell thunderstorm. These storms can live for hours, producing strong winds, large hail, and tornadoes. Photo by Steve Albers. Courtesy of the National Oceanic and Atmospheric Administration (NOAA).

ever, as in the case of the Illinois-Indiana tornado on May 26, 1917, some can exist for several hours and travel hundreds of kilometers. Because of its geographic size and of the air masses that collide over the middle continent, the United States experiences more tornadoes than any other country in the world, averaging more than 800 annually. The majority of these storms occur between March and July.

Tornadoes pack extremely strong winds capable of destroying buildings, uprooting trees, and hurling objects over distances of several kilometers. These powerful storms are one of the least understood meteorological phenomena. Recent records indicate that perhaps we are now seeing more of these extreme events, although these new numbers could very well be related to a growing population living in previously sparsely populated areas. The presence of people, coupled with increased tornado awareness, leads to tornado sightings in areas where there had previously been no one to report them.

In January 1999, 163 tornadoes were reported in the United States, more than three times the previous January record. Meteorologist Harold Brooks at the National Severe Storms Laboratory in Norman, Oklahoma, attributed these unusual January tornadoes to abnormal weather patterns that brought low-level moisture from the Gulf of Mexico. The southerly winds from the gulf combined with strong upper-level winds to create favorable conditions for tornadoes. Such conditions are typically not seen until March or April. The result was an extreme month, with 141 tornadoes striking Arkansas, Louisiana, and Tennessee, including a one-day January record of 87 tornadoes on January 21.

And January was just the beginning. On May 3, 1999, violent tornadoes roared through Oklahoma and Kansas, destroying over 3,700 homes and businesses, damaging over 16,000 others, and killing 48 people. A tornado's destructiveness is rated according to the Fujita-Pearson Tornado Intensity Scale, formulated in 1971. This scale is shown in Table 1.1. In Oklahoma, a tornado with an F5 rating spent almost 90 minutes ripping a 38-mile path through Oklahoma City and other communities just to the south. According to the Fujita scale, an F5 tornado possesses estimated wind speeds of 429 to 512 kilometers (261 to 318 miles) per hour. A tornado of this intensity can lift strong frame houses completely off their foundations and carry them considerable distances to disintegrate. Meteorologists have long considered the total destructive powers of an F5 tornado to be the maximum winds possible. Although the Fujita scale includes ratings of F6 and higher for wind speeds 513 kilometers (319 miles) per

Figure 1.5. Tornado. More than 800 tornadoes occur in the United States annually. They remain one of the least understood meteorological phenomena. Photo by Steve Albers. Courtesy Sky & Telescope.

Table 1.1
Fujita-Pearson Tornado Intensity Scale.

F Number	F-Scale Damage Specification
F0	18–32 m/sec (40–72 mph): Light damage; some damage to chimneys and signboards; branches off trees; shallow-rooted trees pushed over
F1	33–49 m/sec (73–112 mph): Moderate damage; the lower limit (73 mph) is the beginning of hurricane wind speed; surfaces peeled off roofs; mobile homes pushed off foundations or overturned; moving autos pushed off road
F2	50–69 m/sec (113–157 mph): Considerable damage; roofs torn off frame houses; mobile homes demolished; boxcars pushed over; large trees snapped or uprooted; light-object missiles generated
F3	70–92 m/sec (158–206 mph): Severe damage; roofs and some walls torn off well-constructed houses; trains overturned; most trees in forest uprooted; heavy cars lifted off the ground and thrown
F4	93–116 m/sec (207–260 mph): Devastating damage; well-constructed houses leveled; structures with weak foundations blown some distance; cars and other objects thrown as large, high-speed missiles
F5	117–142 m/sec (261–318 mph): Incredible damage; strong frame houses lifted off foundations and carried considerable distance to disintegrate; automobile-sized missiles fly through the air in excess of 100 meters; trees debarked
F6–12	143 m/sec to Mach 1, the speed of sound (319–700 mph): These wind speeds are generally considered to be impossible, although the subject is a topic of debate

Note: m/sec = meters per second

Source: *Tornado Alley: Monster Storms of the Great Plains* by Howard B. Bluestein, copyright © 1999 by Oxford University Press, Inc. Used by permission of Oxford University Press, Inc.

hour or greater, maximum winds in a tornado are not expected to ever reach these levels.

The Oklahoma City tornado was remarkable in several ways. First, meteorologists were able to track the storm from the time it developed until the time it dissipated, giving people advance warning and saving lives. Second, this tornado bore the highest wind speeds ever measured: approximately 500 kilometers (310 miles) per hour, near the upper limit of the F5 range. Joshua Wurman and colleagues at the University of Oklahoma measured these winds using a mobile radar unit that estimates wind based on the speed of rain droplets and debris inside a storm (see Chapter 2). These and other measurements taken from the Oklahoma City tornado provide scientists with an opportunity to learn more about tornado formation and evolution processes.

Such measurements can be difficult to obtain but are extremely valuable. Despite improvements in observation and forecasting abilities, the exact mechanism that triggers tornado formation remains elusive. By discovering this mechanism, meteorologists would be able to improve tornado forecasts significantly. This discovery could also lead to a better understanding of whether a tornado will be small and less threatening, or an F5 super-twister like the one experienced in Oklahoma City.

Current tornado research is aimed primarily at answering the questions above. Howard Bluestein, a researcher at the University of Oklahoma, is a pioneer in this field and has recently published a book describing in great detail his experiences in chasing and studying tornadoes. A large part of recent tornado research efforts has involved field work: chasing and observing actual storms. Scientists use mobile radars and other instruments to try to probe these storms and penetrate their secrets. One such experiment was project VORTEX (Verification of the Origins of Rotation in Tornadoes Experiment), designed as a collaborative effort between the National Severe Storms Laboratory (NSSL) and other research agencies and universities. According to Erik Rasmussen, a researcher at NSSL and VORTEX project director, its primary objective was to understand what causes tornadoes to form. The VORTEX project involved measuring winds, temperature, moisture, and other conditions near a severe storm and then analyzing the data for observable signals that distinguish storms that produce tornadoes from those that do not.

Data gathering for VORTEX was carried out in the central United States during 1994 and 1995. In 1994, VORTEX researchers operated in the field 18 days and were successful in obtaining observations of supercell storms on 9 of them. In 1995, another 18 days of field operations resulted in 13 days of quality data collecting and nine tornado intercepts. The VORTEX data provide a powerful tool for seeing into tornadoes and the thunderstorms responsible for their formation. Using these data, scientists have learned that it is not necessary to collect data from an entire supercell storm to determine the origins of tornado formation. Instead, they can focus on a section of the storm only a few miles in area where conditions are most favorable

Figure 1.6. A Project VORTEX field crew prepares for an encounter with another severe storm. The group will collect measurements to help understand what causes tornadoes to form. Courtesy of the National Oceanic and Atmospheric Administration (NOAA).

to tornado development. This knowledge led to the proposal of a sub-VORTEX project in spring 1997. The project was a smaller version of VORTEX, involving fewer measurement resources but designed to focus more closely on specific areas of the storm. Analysis of the VORTEX and sub-VORTEX data is still in progress. Nevertheless, these observations provide new insights into tornado formation and behavior, which scientists hope to translate into improved tornado forecasts and warnings.

References
Bluestein, H.B. *Tornado Alley: Monster Storms of the Great Plains.* Oxford: Oxford University Press, 1999.
Britt, R.R. "Oklahoma Tornado Yields Fastest Wind Ever Recorded, Plus Clues to Twister Dynamics." explorezone.com, May 12, 1999.
Brooks, H. Tornado information from National Severe Storms Laboratory. Oct. 1, 1998. <http://www.nssl.noaa.gov/~brooks/tornado/#alltorn>.
Doswell, C. "Tornadoes: Some Hard Realities." 1999. <http://www.wildstar.net/~doswell/Tornado-essay.html>.
"Early Tornado Warnings Saved Lives on the Plains." *NOAA Report*, 8(6), June 1999.
"January Sets Tornado Record." CNN Interactive, Jan. 10, 1999.
National Severe Storms Laboratory, <http://www.nssl.noaa.gov>.
Storm Prediction Center, <http://www.spc.noaa.gov>.

"Storm Prediction Makes the Difference, Scientists Say." CNN Interactive, May 6, 1999.

Wurman, J. Personal communication, 2000.

Hurricanes

Hurricanes are severe weather events common to some coastal and island areas in the United States and other parts of the world. Hurricanes are huge, often massively destructive storms that are born over the warm waters of oceans. The United States is influenced directly by hurricanes in the western Atlantic and eastern Pacific basins. Western Pacific hurricanes, which threaten Japan and the Pacific islands, are known as typhoons. In the areas bordering the Indian Ocean, the Bay of Bengal, and the Arabian Sea, these storms are called tropical cyclones. Whatever the name, they can pack winds of 120 to 240 kilometers (75 to 150 miles) an hour or more and bring devastating amounts of rainfall.

Most hurricanes form in the tropical oceans below 20 degrees latitude, where the air is already warm and humid because of heat transfer and evaporation from the warm ocean. As winds from different directions clash over warm water, the moist, warm air is lifted upward. If the resulting thunderstorms organize themselves into a large enough area, the earth's rotation helps to get the winds rotating counterclockwise (clockwise in the Southern Hemisphere) in a large circle. Only a few thunderstorm clusters grow to this large size, and only a few of these slowly rotating tropical depressions, as they are called, intensify into hurricanes. At the center of the tropical depression is a column of calm air around which the thunderstorms swirl. This column is called the *eye*, and the storms directly bordering it, composing the part of the hurricane called the eye wall, pack the highest, most deadly winds.

Hurricanes are classified by intensity into one of five categories on the Saffir-Simpson Hurricane Scale. This classification system can help estimate the potential for flooding and other property damage from winds and storm surge along a coast if a hurricane should make landfall. The scale is based on wind speed: hurricanes having winds of 119 to 153 kilometers (74 to 95 miles) per hour are classified as Category 1. The accompanying storm surge from a Category 1 hurricane is expected to be about 1.2 to 1.5 meters (4 to 5 feet) above normal. Category 5 hurricanes have winds greater than 249 kilometers (155 miles) per hour and can produce a storm surge of more than 5.5

Meteorology Today 23

Figure 1.7. Satellite image of a hurricane. These large storms form over warm ocean water and can cause significant damage from strong winds, heavy rain, storm surges, and flooding. Courtesy of Wilfred von Dauster, National Oceanic and Atmospheric Administration (NOAA).

meters (18 feet) above normal. These hurricanes can cause catastrophic damage, requiring massive evacuations within 8 to 16 kilometers (5 to 10 miles) of a shoreline. Hurricane Gilbert, making its appearance in 1988, was a Category 5 hurricane at peak intensity and is the strongest Atlantic hurricane on record.

The hurricanes most threatening to the United States are those that come near or onshore along the Atlantic seaboard and along the Gulf Coast. The Atlantic hurricane season runs June 1 through November 30 of each year. By the end of the 1998 season, scientists were calling the period from 1995 to 1998 the most active years ever for hurricane activity in the Atlantic basin. It was no wonder, given that 1998 alone saw nine hurricanes and five tropical storms. September 1998 marked the first time since 1893 that four hurricanes (Georges, Ivan, Jeanne, and Karl) were in simultaneous existence in the Atlantic basin. Some of the most vigorous storms of the 1998 hurricane season were:

- Bonnie (August 19–30), causing $1 billion damage in North Carolina
- Tropical storm Charley, which dumped over 50 centimeters (18 inches) of rain near Del Rio, Texas, on August 21 and 22
- Earl (August 31–September 3), responsible for $25 million in losses and three deaths in the Florida Panhandle
- Tropical storm Frances (September 8–11), which killed one person and flooded more than 300 miles of the Gulf Coast from Texas to Louisiana
- Georges (September 15–19), a Category 4 storm with winds of 240 kilometers (150 miles) per hour, contributing to over 350 deaths in Haiti and the Dominican Republic
- Mitch (October 22–November 6), a Category 5 storm with sustained winds of 250 kilometers (155 miles) per hour and torrential rains, causing flooding and over 10,000 deaths in Central America

In all, damage from Hurricane Georges's destructive path along the Caribbean and four Gulf Coast states was estimated at $2 billion. Hurricane Mitch ranks as the third deadliest hurricane on record. It is little comfort to the victims in the devastated areas that most of the death and destruction from floods came after Mitch had moved inland and weakened into a tropical depression.

Even near the start of the 1998 hurricane season, scientists were theorizing that we are entering a cycle of more violent, more frequent storms. Recent research has been aimed toward a phenomenon known as the "Atlantic conveyor." This conveyor transports cold water from the seas of Iceland and Greenland along the bottom of the ocean to Antarctica. After several decades, this water surfaces and flows back northward, warming along its course across the equator. According to hurricane expert William Gray at Colorado State University, the Atlantic conveyor may have picked up speed lately, moving the warm equatorial water north to midlatitudes before it has much time to cool. The warm water provides energy for hurricanes. *La Niña*, the name for climate conditions associated with a cooling

Meteorology Today

Figure 1.8. The cockpit of a U.S. Air Force Lockheed WC-130 "Hercules" Hurricane Hunter airplane. Photo by Scott Dommin.

of the central and eastern tropical Pacific, can also affect hurricane development. In La Niña years, upper-level winds tend to be easterly, similar to the easterly winds near the ocean surface. In contrast to the upper-level westerlies typical of El Niño years, this smaller contrast between low- and high-level winds can favor the birth of hurricanes.

The 1998 hurricane season provided scientists many opportunities to study these storms up close and personal. Hurricanes are beasts of an entirely different nature from the generally well understood air mass thunderstorms. These massive storms attract the attention of scientists desiring insight into their formation and behavior. And yes, there are "Hurricane Hunters." They are the 53rd Weather Reconnaissance Squadron of the Air Force Reserve and have been flying into hurricanes since 1944. The National Hurricane Center of the National Oceanic and Atmospheric Administration also flies weather reconnaissance aircraft.

Hurricane investigators view their targets from satellites, aircraft, and radar and other ground-based measurement systems. What they find is sometimes surprising. Scientists studying Hurricane Bonnie in 1998 were astonished to find that the outflow west of the hurricane's eye was not being generated from the west side of the eye wall but from an immense vertically developed cloud far away on the east side of the eye. Ed Zipser, a hurricane expert at Texas A&M (now at the University of Utah), was on board a research plane that flew under this outflow at 11,000 meters (35,000 feet). By flying through and above Bonnie and other hurricanes, Zipser and colleagues collect valuable data for improving the understanding of hurricanes. What they learn can help prepare coastal residents and ultimately save lives.

When the hurricane season started in 1999, researchers were ready to go. On May 27, 1999, the National Oceanic and Atmospheric Administration's Climate Prediction Center (CPC) and National Hurricane Center (NHC) released their outlook for the June to November season. The outlook indicated a strong likelihood of above-average tropical storm and hurricane activity over the North Atlantic basin, including at least three major hurricanes. Their forecast proved to be accurate. Tropical storm Arlene threatened Bermuda during the early part of the hurricane season (June 11–18). By August, the ocean waters were warmed up and ready, bringing on Hurricane Bret (August 18–23), a Category 4 storm making landfall in a remote area between Brownsville and Corpus Christi, Texas. Bret was followed by Hurricanes Cindy (August 19–31) and Dennis (August 24–September 5) and tropical storm Emily (August 24–28).

By far the biggest storm of the 1999 Atlantic hurricane season was Floyd (September 7–17). Called "Furious" Floyd, this Category 4 storm was a monster, covering an area over 600 miles and threatening the Atlantic coast with sustained winds of almost 250 kilometers (155 miles) per hour. Along the Florida, Georgia, and North and South Carolina coastlines, emergency response officials ordered 3 million people to evacuate—the largest number in history. Floyd took forty-one lives and caused heavy rain and flooding up and down the Atlantic coast, from Florida all the way north to Maine. The worst flooding by far occurred in North Carolina.

Hurricane Gert (September 11–23) brought damaging winds to Bermuda and was followed by tropical storm Harvey (September 19–22), which dumped over 10 inches of rain on some communities in southwest Florida. Hurricane Irene (October 13–19) was a Category 1 storm able to cause significant damage in Florida and more flooding in North Carolina, a state already hit hard by both Dennis and Floyd. Hurricane José (October 17–25) hit the Lesser Antilles and Puerto Rico with rain and wind, and tropical storm Katrina (October 28–November 1) made landfall in Central America, where it was soon downgraded to a tropical depression. Hurricane Lenny (November

13–20), a Category 4 storm, was the strongest Atlantic hurricane on record to occur so late in the season, and was also unusual because of its eastward motion across the Caribbean. In total, the 1999 Atlantic hurricane season brought the formation of twelve tropical storms, eight of which became hurricanes, including five that were classified as major, having wind speeds of Category 3 or above on the Saffir-Simpson Scale.

Hurricanes are not confined to the Atlantic Ocean. In early November 1999, the Indian Ocean version of a hurricane, called a tropical cyclone, devastated large areas of India's coastline. By November 7, the death toll was estimated at between 1,715 and 2,464 persons or more. The cyclone caused so much flooding and so many deaths of people and livestock in eastern India that corpses could not be attended to properly. The water supply became contaminated, and diseases such as gastroenteritis spread quickly, increasing the death toll. Officials say the damage could have been much worse: twenty-three cyclone shelters built by the Red Cross were credited with saving 34,500 lives.

The hurricanes of 1999, responsible for flooding, property damage, and deaths, may have lingering effects in other ways. Jerry Jarrell, director of the National Hurricane Center in the United States, and other researchers are working to develop a new scale for evaluating these powerful storms. The proposed scale would account for the potential of rainfall, flooding, tornadoes, and other impacts associated with a hurricane. The scale would complement the Saffir-Simpson Scale, which is based on sustained wind speeds, and would help to inform the public of a hurricane's potential for destruction. This rating could be used to help emergency authorities and coastal residents better prepare for a storm. The new scale requires approval by the National Oceanic and Atmospheric Adminstration and other agencies.

References

Bove, M.C., et al. "Effect of El Niño on U.S. Landfalling Hurricanes, Revisited." *Bulletin of the American Meteorological Society*, 79(11): 2477–2482, Nov. 1998.

"Busy Atlantic-Caribbean Hurricane Season Predicted." CNN Interactive, Apr. 7, 1999.

Chartuk, R. "Floyd Floods U.S. East Coast." *NOAA Report*, 8(9), Oct. 1999.

Climate Prediction Center, <http://www.cpc.noaa.gov>.

Hager, R. "Understanding Monster Storms." MSNBC, Aug. 23, 1999.

Kerr, R.A. "Forecasters Learning to Read a Hurricane's Mind." *Science*, 284: 563–565, 1999.

Larson, E. "Waiting for Hurricane X." *Time*, Sept. 7, 1998, pp. 62–66.
Lemonick, M. "A Very Close Call." *Time*, Sept. 27, 1999, pp. 34–37.
Lockridge, R. "Forecasters Employ New Technology to Get Jump on Hurricanes." CNN Interactive, Aug. 22, 1999.
"Mitch May Have Blown Away U.S. Hurricane Scale." CNN Interactive, Nov. 23, 1998.
Nash, J.M. "Wait Till Next Time." *Time*, Sept. 27, 1999, pp. 39–40.
National Centers for Environmental Prediction, <http://www.ncep.noaa.gov/>.
National Hurricane Center, <http://www.nhc.noaa.gov/>.
Zipser, E. Personal communication, 2000.

Heat, Drought, and Other Weather Extremes

Severe or extreme weather events also include heat waves, drought, floods, winterstorms, and blizzards. Heat waves can lead to an excessive loss of life and are often coincident with drought incidents, which are particularly damaging to agriculture. In August 1999, the National Oceanic and Atmospheric Administration (NOAA) began issuing weekly heat threat forecasts, giving managers, planners, and the public a better chance to prepare for heat waves. These forecasts will eventually be extended to up to two weeks in advance and provide probabilities of the number of consecutive days the heat index (a value combining the effects of temperature and humidity) will exceed critical values. In conjunction with the Department of Agriculture and the National Drought Mitigation Center, NOAA also offers a drought monitoring service to provide information on the extent and intensity of drought conditions nationwide. This information can be critical to agriculture, water management, and fire weather applications.

The societal impacts of drought and floods are discussed further in Chapter 3. The recent advances in weather forecasting, highlighted in the sections above, can assist emergency managers, highway departments, and the public to be better prepared and informed. This information helps save lives and property and can affect all facets of society, from economics to agriculture to industry and construction.

References
National Climatic Data Center, <http://www.ncdc.noaa.gov>.
National Drought Mitigation Center, <http://enso.unl.edu/ndmc>.
National Oceanic and Atmospheric Administration, <http://www.noaa.gov>.
National Weather Service Drought Page, <http://www.nws.noaa.gov/om/drought.html>.

EL NIÑO AND LA NIÑA

To anyone watching the news in 1997 and 1998, it may have seemed as if some mysterious entity called El Niño was running around wreaking havoc on the planet. El Niño was blamed for wacky winter weather in many parts of the world and was labeled responsible for increased rainfall, flooding, and tornadoes in many parts of the United States. El Niño was said to have influenced world grain markets and oil prices, and even job growth. As one example, consider that above-normal precipitation in usually dry areas such as the southwestern United States could lead to increased employment and profits in construction as homeowners repair leaky roofs. El Niño was blamed for the January 1998 ice storm affecting Canada and the northeastern United States and for droughts in Texas. But in all this discussion, what really is El Niño, and can it actually be linked to all of this?

Many of the events occurring in 1997 and 1998 had nothing at all to do with El Niño. El Niño conditions can be linked to intensified storm activity, and hence increased rainfall, in California and the southern states and along the west coast of South America. An El Niño year may also correspond to above-normal winter temperatures coupled with below-normal precipitation and snowfall in the northern part of the United States. On a world scale, El Niño can dramatically affect fishing industries off the west coasts of North and South America and may be associated with droughts in Indonesia, Africa, and Australia.

El Niño is an anomalous, or irregular, warming of surface waters in the eastern tropical Pacific Ocean. This area of warm water can be up to one and a half times the size of the United States and tends to occur every two to seven years off the west coast of Peru. These conditions often become evident in the later part of the year (around Christmas)—hence, the name *El Niño*, which is Spanish for "boy child." The conditions may persist for a few weeks up to a period of a couple years.

El Niño is linked with an atmospheric phenomenon referred to as the Southern Oscillation, usually written as the El Niño/Southern Oscillation (ENSO). This term refers to ocean-atmosphere interactions involving large-scale swings in air pressure between the western and eastern tropical Pacific. Normally, east-to-west winds (called trade winds) over the Pacific Ocean push water from the South American coast toward Indonesia. The pressure differences associated with ENSO can weaken or even reverse these trade winds, causing warm surface water located off the coast of Australia to expand eastward into the central and eastern Pacific. Warming of the surface water in these areas suppresses the normal cold water upwelling of the coasts of northern Peru and Ecuador. The resulting El Niño conditions can affect weather in many widely separated regions of the globe.

Although there is some debate concerning the number and duration of previous El Niño events, the El Niño of 1997–1998 is generally considered the twenty-fifth such event to occur during the 1900s. However, this El Niño was immensely different from its predecessors in two important ways. First, scientists were able to predict the start and scope of this El Niño, even six months ahead of time. Second, the event was forecast to be the strongest ever. As a result of these predictions, economic concerns and public awareness were running high, making this the most monitored and discussed El Niño event in history. A special El Niño symposium was held in Boulder, Colorado, to discuss El Niño effects, the Federal Emergency Management Agency (FEMA) held its own El Niño summit, and $18 million was spent to establish an international El Niño research center in New York.

The 1997–1998 El Niño was studied by instruments on buoys, on ships, and aboard satellites and was simulated and analyzed using models run on supercomputers. One of these ocean-atmosphere measurement programs, the Tropical Ocean Global Atmosphere-Tropical Atmosphere-Ocean (TOGA-TAO) Program, was established in the aftermath of 2,000 deaths and $13 billion in losses associated with the 1982–1983 El Niño. This international project placed some seventy buoys across the equatorial Pacific to provide important oceanic and atmospheric data to forecasters worldwide. These and other observations were combined with improved seasonal forecast models to predict the early onset of an ENSO shift in the spring of 1997 and particularly strong El Niño conditions during the 1997–1998 winter. These forecasts were produced by the National Centers for Environmental Prediction (NCEP) in the United States, as well as by at least ten other research institutes worldwide, and provided advance information of El Niño and its potential impacts, including heavy winter rainfall in California and the Gulf Coast region. Other climate effects of the 1997–1998 El Niño were less accurately predicted, especially the extreme conditions that plagued the southeastern United States with drought during the spring of 1998 but dumped damaging amounts of rain on Georgia crops. One factor in the less-than-good forecasts during this time could be associated with a temporary break in El Niño conditions during March 1998.

Forecasting El Niño remains a difficult process. To date, a reliable long-range El Niño forecast system has not been developed. Stephen Zebiak of the International Research Institute/Lamont-Doherty Earth Observatory has pointed out a number of limits that affect the production of long-range El Niño forecasts. These limits include not only problems with the models and the way data are used in the models, but also gaps in the observing system and on the inherent unpredictability of the climate system.

Most scientists agree that the 1997–1998 El Niño was one of the strongest on record. The media, very much caught up in the El Niño hoopla, called it "The Climate Event of the Century." And despite the lingering difficulties of El Niño forecasting, the ability to predict this El Niño successfully was a dramatic improvement over scientists' abilities just a decade and a half or so ago, when the 1982–1983 El Niño was not even detected until it was nearly at its peak. Another success of the 1997–1998 forecasts was the ability of many models to estimate that the end of El Niño, followed by the development of La Niña (cooler surface water) conditions, would occur sometime during the second half of 1998. These forecasts did not predict exactly when conditions would change from an El Niño to a La Niña state or how quickly this transition would occur. However, the general agreement among the forecast models that the switch would occur within a given time frame (and a time frame that was close to what was actually observed) was considered a significant advancement in climate modeling.

The strength and duration of the 1997–1998 El Niño episode also caught the attention of researchers concerned with the effects of global climate change. Scientists have speculated that increased greenhouse warming (discussed in later in this chapter) could result in more frequent, more extreme El Niño events. Current reports indicate that over the past fifty or so years, and especially during the past twenty-five years, there has been a gradual tilt toward more El Niño events, accompanied by a reduction in the number of La Niña events. Scientists generally concur that there have been from two to four La Niña episodes since 1970. The number of El Niño events is considered to be twice this. Also, the warming of the central and eastern Pacific waters can provide a good view of what life in a warmer world would be like. The difference, of course, is that temperature differences associated with global warming may be realized in more locations than a single portion of an ocean basin.

El Niño has a sister, *La Niña*, from the Spanish word for "girl child." La Niña is the antithesis of El Niño and refers to a cooling of sea surface temperatures in the central and eastern tropical Pacific (the area warmed during El Niño). La Niña is generally associated with strong trade winds, which push the warm surface water of El Niño westward and allow strong upwelling of cold, deep ocean water along the west coasts of North and South America. James O'Brien of Florida State University points out that in a strict sense, La Niña is defined as an extreme cooling, as opposed to just any cooling, of the central equatorial Pacific waters. La Niña is not simply the retreat of El Niño. In fact, not all El Niño events are followed by subsequent La Niña occurrences.

The most recent La Niña episode began in late May and June 1998. According to Michael McPhaden of the National Oceanic and Atmospheric Laboratory's Pacific Marine Environmental Laboratory, this period witnessed a drop in ocean surface—temperatures dropped by 8°C (10°F)—truly indicating the end of El Niño. By March 1999, the NCEP Climate Prediction Center was calling this La Niña one of the strongest of the past fifty years.

In terms of its effects on climate, La Niña is not simply the opposite of El Niño. But in many cases, the weather conditions spawned by El Niño's warm pool will be suppressed by La Niña's cooling effect, and vice versa. The effects of La Niña tend to be most pronounced in the winter and generally result in below-normal temperatures in the Central Plains and Midwest, unusually wet weather in the Northeast and Pacific Northwest, and warmer and drier conditions in the Southeast. Like her brother, La Niña is often held accountable for any anomalous or extreme weather events that occur during her presence in the Pacific. The major drought affecting the Midwest in 1988 has often been attributed to La Niña conditions during 1988–1989. However, as researchers identify other climate scenarios capable of producing droughts of this magnitude, La Niña's role in causing these conditions becomes considerably less substantiated.

Together, El Niño and La Niña represent extremes of the Southern Oscillation. As such, many of the trials and tribulations associated with predicting El Niño and its effects also apply to La Niña. Researchers suggest that the La Niña may in fact be even more difficult to predict than El Niño. Buoy and satellite instruments remain in place to monitor the surface water conditions and report the observations to scientists in laboratories and research institutes worldwide. These scientists look for ways to improve forecasts of El Niño, La Niña, and associated effects and to improve public understanding of these effects, not only at regional levels but also in terms of local decision making.

References

CNN Interactive, <http://cnn.com/SPECIALS/el.nino/>.
El Niño Theme Page, <http://www.pmel.noaa.gov/toga-tao/el-nino/nino-home.html>.
Glantz, M. *Currents of Change: El Niño's Impact on Climate and Society.* Cambridge: Cambridge University Press, 1996.
Global Effects Page, <http://www1.tor.ec.gc/elnino/global/index_e.cfm>.
Kerr, R.A. "Big El Niños Ride the Back of Slower Climate Change." *Science*, 283: 1108–1109, 1999.
La Niña Resources, <http://www1.tor.ec.gc.ca/lanina/further_reading/index_e.cfm>.
La Niña Summit, Boulder, Colorado, June 1998. Report available on-line at <http://www.esig.ucar.edu>.
National Aeronautics and Space Administration, <http://www.nasa.gov>.
National Oceanic and Atmospheric Administration, <http://www.noaa.gov>.
Neelin, J.D., and M. Latif. "El Niño Dynamics." *Physics Today*, 51(12): 32–36, Dec. 1998.
Pacific Marine Environmental Laboratories, <http://www.pmel.noaa.gov>.
Seabrook, C. "Weather Wiser: New Technology Stretched Across Pacific Beefs Up Knowledge of El Niño, Sister." *Atlanta Journal-Constitution*, Oct. 25, 1998.
"Strange Brew: Bringer of Bounty and Famine, El Niño Keeps Experts Guessing." CNN Interactive, 1998.
Timmermann, A., J. Oberhuber, A. Bacher, M. Esch, M. Latif, and E. Roeckner. "Increased El Niño Frequency in a Climate Model Forced by Future Greenhouse Warming." *Nature*, 398:694–696, 1999.

ATMOSPHERIC CHEMISTRY

The previous sections of this chapter have focused mainly on the dynamics, or motions, of the atmosphere and their role in influencing weather and climate events such as severe storms and the El Niño/Southern Oscillation. Although atmospheric dynamics is certainly a part of and connected to atmospheric chemistry, we are now entering a realm where concern is focused less on atmosphere motions and on the forces influencing them and more on the composition of the atmosphere and the interactions of these constituents. This is the realm of atmospheric chemistry, a discipline that originally evolved not from meteorology but from chemistry.

Chemistry, in its modern and quantitative form, had its beginnings in the eighteenth century. It was early chemists such as Joseph Priestley, Antoine-Laurent Lavoisier, and Henry Cavendish who first identified and quantified the chemical components of the atmosphere, including oxygen, nitrogen, water vapor, and carbon dioxide. In the nineteenth century, attention shifted away from the major atmospheric constituents toward studies of trace gases and atmospheric particulate (or aerosol) species. Special concern was focused on understanding the impacts of these trace constituents on the environment. The rise of potentially critical environmental problems in the latter half of the twentieth century greatly expanded the role of atmospheric chemistry in society. Suddenly, atmospheric chemistry research was relevant to a number of policy issues. Suddenly, the implications of the species composing our atmosphere, and in particular, any changes in the quantities of these species, could be documented, and were found to influence us all.

Stratospheric Ozone

Ozone is a very small part of the blanket of air that envelops our planet. Only about one molecule out of every million oxygen molecules in the atmosphere is in the form of ozone. While the more abundant molecular oxygen contains two oxygen atoms, ozone is composed of three. Most of the earth's ozone is located in a layer in the lower stratosphere, between 20 and 30 kilometers (12 to 19 miles) altitude. This ozone is essential to the health of life on our planet. By absorbing ultraviolet radiation from the sun, ozone prevents a portion of these rays from reaching the earth's surface. In increased amounts, this ultraviolet radiation can be harmful to plants and animals.

Ozone has received a lot of attention in recent years. In 1995, Mario Molina, Sherwood Rowland, and Paul Crutzen shared the Nobel Prize for chemistry for their work linking chemical reactions and the depletion of the ozone layer. Molina and Rowland had proffered their theories on chlorofluorocarbons and ozone loss some twenty years earlier. Crutzen's work, also done in the 1970s, addressed the role of nitrogen oxides, such as those emitted by supersonic aircraft, in ozone depletion. Ozone was the first atmospheric compound to be considered an endangered species and the first chemical to be preserved by international treaty.

The attention was spurred by observations of rather significant decreases in the amount of ozone in the earth's stratosphere. These decreases were first observed by British Antarctic Survey scientists Joseph Farman, B.G. Gardiner, and J.D. Shanklin, who had been making annual trips to Antarctica for more than ten years. In October 1984, their measurements of the total column ozone over Halley Bay, close to the South Pole, showed alarming decreases. Ten years prior to this, Sherwood Rowland and Mario Molina had put forth their theory that commonly used chemicals, called chlorofluorocarbons (CFCs), could accumulate in the stratosphere and lead to ozone loss. Until the British Antarctic Survey report, no direct observations had been documented to verify this theory. Even with the direct observations, it would still require a few more years to verify the theory and gain a better understanding of the ozone production and destruction processes in the stratosphere, particularly in the polar regions.

CFCs were first developed in the 1930s for use as refrigerants, cleaning solvents, and aerosol spray propellants. The most commonly manufactured CFCs have been chlorofluorocarbon-11 ($CFCl_3$) and chlorofluorocarbon-12 (CF_2Cl_2). Because these compounds are chemically very stable, their popularity, as well as the realm of potential uses, grew. CFC production increased exponentially until 1974, when Ralph Cicerone and Richard Stolarski, then at the University of Michigan, proposed that chlorine may play a role in stratospheric ozone destruction. The link between CFCs and ozone depletion was put forth by Mario Molina and Sherwood Rowland, at that time at the University of California at Irvine. The public and politicians listened, and in 1978, CFC-propelled hair sprays and deodorants were banned. But without verification of Molina and Rowland's theory and with heavy influence from CFC industries, further restrictions on CFC production were not possible. By 1985, the annual production of CFCs rivaled the levels seen in the early 1970s.

Ozone quantities are typically given as a total column amount, representing the total number of ozone molecules in an imaginary tube 1 centimeter (0.4 inch) on a side, stretching upward from the surface to the top of the atmosphere. Atmospheric scientists typically express this amount in Dobson Units (DU), named after George Dobson (1889–1976), an English physicist and pioneer in atmospheric ozone measurements. One DU represents 2.7×10^{16} ozone molecules. Typical column ozone amounts are about 300 DU, or roughly 800×10^{16} ozone molecules. If compressed to a layer of pure ozone at 1-atmosphere pressure, these molecules would form a layer only 3 millimeters (0.1 inch) thick. Nevertheless, these small amounts of ozone are crucially important to life on earth.

The ozone losses measured by Farman and colleagues in Antarctica in 1984 were largest during the Antarctic springtime. Measurements showed that the total column ozone for October had decreased by an order of about 40 percent, from over 340 DU to less than 200 DU.

In 1986, the United States sent the first National Ozone Expedition (NOZE 1) to Antarctica to research ozone depletion. Susan Solomon of the National Oceanic and Atmospheric Administration's Aeronomy Laboratory led the sixteen-member science team in making measurements of trace gases and the physical properties of the atmosphere. Their work, in collaboration with scientists from the National Aeronautics and Space Administration and other organizations, provided evidence that CFC ozone depletion could be accelerated by the presence of polar stratospheric clouds during Antarctica winters. Polar stratospheric clouds (PSCs) form high in the atmosphere above the South Pole during the cold, dark winters. The clouds are composed of either ice particles or a combination of ice particles and frozen nitric acid. These particle surfaces enhance the chemical transformation of gaseous chlorine compounds, originating from CFCs, into forms that can catalytically destroy ozone when sunrise over Antarctica begins. By comparing laboratory knowledge and theory with the results observed in the field, scientists found that the decreases in ozone levels were consistent with the results predicted from the chemical reactions. By linking PSCs to chemical reactions involving chlorine, the knowledge of ozone destruction mechanisms in the Antarctic was largely complete.

By 1990, ozone amounts had dropped to 120 DU, or about one-third of the ozone amounts present in the 1960s. Satellite monitoring also showed that the area of ozone depletion, referred to as the *ozone hole* and indicating ozone amounts less than 220 DU, had increased in size, covering almost the entire Antarctic continent. Low ozone amounts have been observed over Australia and southern Chile as well. In 1998, Southern Hemisphere ozone amounts were less than 100 DU, and the areal extent of the ozone hole was a record 17 million square kilometers (6.5 million square miles). In September 2000 (spring in the Southern Hemisphere), this record was surpassed, with the areal extent of the ozone hole covering 28.3 million square kilometers (11 million square miles). Total column ozone values measured over the South Pole in 1999 are shown in Figure 1.9. The observations indicated that the area of the ozone hole was slightly

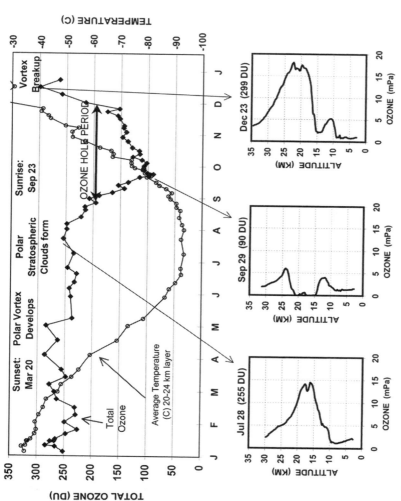

Figure 1.9. Total ozone measurements over the South Pole in 1999, marked by diamond symbols on the solid line. The open circles indicate average temperatures at 20–24 kilometers. The temperatures reach a minimum in July and August, corresponding to the Southern Hemisphere winter. The cold temperatures cause polar stratospheric clouds to form; chemical reactions on these clouds accelerate ozone destruction. In late September, the total ozone amount fell to less than 100 DU. As indicated by the vertical profiles, ozone was almost entirely depleted at altitudes between 15 and 17 km. Figures provided by B. Johnson, NOAA/CMDL.

smaller than the record set the year before, but the total column ozone values were again down to approximately 90 DU.

In addition, ozone losses on the order of 20 to 25 percent were observed in the Arctic during six of the nine years from 1990 to 1998. Arctic ozone values are highly dependent on dynamic conditions. The Arctic stratosphere is more unstable than that over the Antarctic and also tends to be warmer. Fewer polar stratospheric clouds and the greater chance of mixing in ozone-rich air from midlatitudes mean that severe ozone destruction in the northern high latitudes may be restrained. This does not mean that Antarctic ozone depletion levels are not possible in the Arctic. According to Bryan Johnson of the National Oceanic and Atmospheric Administration's (NOAA's) Climate Monitoring and Diagnostics Laboratory, "There is always the potential for the Arctic to have near-Antarctic-type ozone depletion. We just need one good cold stable stratosphere during the winter and early springtime, with little mixing from northern mid-latitudes." In the winter and spring of 2000, these conditions occurred, and Arctic ozone levels showed significant decreases—as much as 60 percent at 18 kilometers (11 miles) altitude.

Ozone losses have also been observed at northern midlatitudes. Column amounts of ozone over the Seattle area, for instance, dropped from 391 DU in 1979 to 360 DU in 1994. Over Los Angeles, the amounts fell from 368 DU (1979) to 330 DU in 1994. As the amount of total column ozone decreases, the amount of ultraviolet radiation from the sun increases. Combined with current outdoor activities and lifestyles, this puts large numbers of people at greater risk for skin cancer, eye disorders, and other health problems. The risks of overexposure to ultraviolet radiation in sunlight are further examined in Chapter 3.

The changes in stratospheric ozone amounts, with accompanying reports of increased incidences of skin cancer and eye disorders linked to ultraviolet radiation exposure, again attracted public and political interest. In 1987, the Montreal Protocol on Substances That Deplete the Ozone Layer was proposed as a framework for phasing out production of CFCs and similar compounds. In 1990 in London and again in 1992 in Copenhagen, amendments were added to accelerate the phase-out and broaden the list of affected chemicals. Following the Copenhagen meeting, the treaty had already been ratified by over fifty countries. By 1996, the production of CFCs and other ozone-depleting substances such as methyl chloroform and carbon tetrachloride was to be completely stopped by the European Union member nations, the United States, and several other nations.

Recent measurements indicate that these reductions are working. A 1999 paper by Stephen Montzka and coauthors at NOAA's Climate Monitoring and Diagnostics Laboratory reports that lower-atmosphere concentrations of ozone-depleting substances, some of which have lifetimes of up to 380 years, have declined about 3 percent overall from 1994. This is mainly due to decreases in methyl chloroform, which has an atmospheric lifetime of 6.5 years. Their measurements also show a decrease in the atmospheric load of trichloroethane (CH_3CCl_3), previously a common cleaning solvent. These decreases can be sustained only through reductions in the use of chemical compounds that lead to ozone depletion. Yet despite international regulation, the global amounts of a few chemicals, including CFC-12 and the bromine compound halon-1211, continue to increase. These chemicals are used in refrigeration and foams; one of them, halon-1211, is particularly harmful in the stratosphere. In the stratosphere, halon-1211 and other bromine compounds can destroy three to fifty times more ozone than would be depleted by chlorine from CFCs.

The accelerated reductions mandated by the London and Copenhagen amendments mean that with international compliance, the ozone layer should soon begin to recover, albeit slowly. Current predictions are for a significant recovery of stratospheric ozone levels by 2050. For many scientists involved in ozone research, this significant recovery will not be a recovery to pre-1970 values. The atmosphere itself is expected to be different by 2050; effects of increasing greenhouse gases on the stratosphere and, correspondingly, on ozone amounts are difficult to predict fully. Other factors complicating ozone recovery are the effects of bromine chemistry and noncompliance with the Montreal Protocol by several industrially expanding countries. Climate-related components, including possible stratospheric cooling linked to surface global warming, complicate the extent and timing of recovery.

In 2000, Elizabeth Weatherhead of the University of Colorado, along with colleagues from several universities and agencies, published a paper reporting that recovery of total column ozone amounts may not be detected for several decades. The length of time is based on difficulties detecting statistically significant changes when analyzing trends in naturally varying geophysical data. Other signs of recovery, including a reduction in stratospheric chlorine loading or increases in ozone amounts at particular layers of the atmosphere, could possibly be detected earlier and are topics of further research.

In general, the late 1990s and early years of the twenty-first century are thought to set the threshold for how low stratospheric ozone levels will dip given current CFC amounts. The scientific consensus is that we are entering a phase of recovery but that this recovery will be long and slow. It is therefore likely that the earth's population will be living under elevated ultraviolet radiation levels for some time to come.

References

Graedel, T.E., and P.J. Crutzen. *Atmospheric Change: An Earth System Perspective.* New York: W.H. Freeman, 1993.

Johnson, B.J. Personal communication, 2000.

Montzka, S.A., J.H. Butler, J.W. Elkins, T.M. Thompson, A.D. Clarke, and L.T. Lock. "Present and Future Trends in the Atmospheric Burden of Ozone-Depleting Halogens." *Nature*, 398: 690–694, 1999.

Turco, R.P. *Earth Under Siege: From Air Pollution to Global Change.* Oxford: Oxford University Press, 1997.

Weatherhead, E.C., et al. "Detecting the Recovery of Total Column Ozone." *Journal of Geophysical Research*, 105(D17): 22, 201–22, 2000.

Weatherhead, E.C. Personal communication, 2000.

World Meteorological Organization. *1998 Scientific Assessment of Ozone Depletion.* Geneva, Switzerland, 1999.

Tropospheric Pollution Issues

While depleting ozone amounts present a problem in the stratosphere, the story in the troposphere is markedly different. The amounts of ozone and other gases in the lower atmosphere have increased dramatically since the beginning of the Industrial Revolution. In the lower atmosphere, these gases are produced directly or indirectly from emissions from power plants, factories, automobiles, and other sources and can influence the atmosphere and overall climate of a region. The effects include degradations in air quality, acid precipitation, and regional cooling due to haze or the presence of particulate matter from dust, soot, or smoke. Tropospheric pollution problems are experienced in every part of the world, although people in some areas are certainly less well off than others when it comes to breathing clean air and seeing a blue sky. Research efforts focused on tropospheric pollution are described in more detail below.

Acid Rain

Emission of air-polluting effluents, in gaseous or particulate form, has a measurable effect on local and regional climate and can even affect the chemical composition of rainfall. During the 1960s, fishermen began reporting that the number and variety of fish in remote lakes of Scandinavia, North America, Scotland, and elsewhere had declined dramatically. Studies of these lakes indicated they had become too acidic for many species to survive. After much more study, scientists had determined the source of the problem: acid rain.

Most fossil fuels contain small amounts of sulfur and nitrogen as impurities. Combustion of these fuels, whether in power plants, internal combustion engines, or for industrial purposes, releases millions of tons of gaseous sulfur dioxide into the atmosphere each year. By the early 1990s, concentrations of atmospheric sulfur in industrial regions of Europe and the United States were ten to twenty times higher than those recorded before coal was burned.

In the atmosphere, sulfur dioxide combines with water vapor to form sulfuric acid. Sulfuric acid can reach the earth's surface as acid rain or snow or as a gaseous or solid sulfate particle. It is responsible for the corrosion of stone, concrete, paint, and steel and the inhibition of germination, plant growth, and nitrogen fixation in the soil. Inhalation of sulfates and dilute concentrations of sulfuric acid can fatally damage the human respiratory system.

International approaches to reduce acid rain were not considered until the early 1980s, when trees began dying in Germany's Black Forest. A 1982 report estimated that 5 percent of the trees had been damaged; by 1985, this figure had risen to 50 percent. Similar observations were made elsewhere in Europe and in the eastern United States. Since the 1980s, sulfur emission from fossil fuel combustion has been on the decrease.

Although scientists discovered early on that acid rain contains nitric as well as sulfuric acid, the nitric acid problem received less attention for many years. Nitrogen-based fertilizers were applied to fields every year to help plants; nitrogen-based acid rain was thought to at least do enough good to balance any potential harm to a forest or field. This was not the case, however, as acids accumulated in snow all winter, only to be released en masse during the spring thaw. This "acid pulse," of which nitric acid is a large component, was responsible for significant damage to lake ecosystems. Nitric acid has been found to contribute to more than 50 percent of the total acidity of rainwater, especially in the western United States, where sulfur from coal is a lesser problem than elsewhere.

Sulfur dioxide and nitrogen oxide emissions are regulated by Title IV of the Clean Air Act Amendments (CAAA) of 1990. These regulations set goals for the reduction of annual sulfur dioxide emissions by electric utilities by an amount of 10 million tons from 1990 levels and for the reduction of nitrogen oxide emissions by 2 million tons. Beginning in 2000, total utility sulfur dioxide emissions will be limited to 8.9 million tons. A 1995 study published by the U.S. Geophysical Survey reports that Phase 1 of the Title IV regulations had led to lower sulfate concentrations in precipitation in the eastern United States, particularly in the Ohio River Valley and the mid-Atlantic region. Nitrate concentrations were not affected.

The trends and effects of acid precipitation are monitored by many agencies and programs, including the U.S. EPA Acid Rain Program, the National Acid Precipitation Assessment Program, and the National Atmospheric Deposition Program/National Trends Network. Canada and many other countries affected by acid precipitation have their own agencies dedicated to researching and monitoring this problem. One example is the RAIN (Reversing Acidification in Norway) project, in which precipitation is collected on a transparent roof to be treated and mixed with sea salts and then reapplied as clean water on the soil below. Between 1984 and 1992, results indicated a decrease in sulfate by almost a factor of three and a reduction in nitrate by more than a factor of six. Sweden, Denmark, the Netherlands, and the Czech Republic have also devoted resources to assessing and controlling the effects of acid rain. Studies in these countries are attempting to understand the effects of acid rain on mountain lakes and watersheds, on pine, spruce, oak, and other trees and vegetation.

References

Firor, J. *The Changing Atmosphere: A Global Challenge*. New Haven, CT: Yale University Press, 1992.

Krecek, J. "Effects of Acid Atmospheric Deposition on Watersheds in Central Europe." In J. Krecek, G.S. Rajwar, and M.J. Haigh (eds.), *Hydrological Problems and Environmental Management in Highlands and Headwaters*. Lisse, The Netherlands: A.A. Balkema, 1996.

Lynch, J.A., V.C. Bowersox, and J.W. Grimm. *Trends in Precipitation Chemistry in the United States, 1983–1994*. Clean Air Act Amendments 1990. Washington, DC: 101st Congress of the United States of America, Jan. 23, 1990.

Skjelkvale, B.L., and R.F. Wright. "Mountain Lakes: Sensitivity to Acid Deposition and Global Climate Change." *Ambio*, 27(4): 280–286, 1998.

van der Salm, C., W. de Vries, M. Olsson, and K. Raulund-Rasmussen. "Modelling Impacts of Atmospheric Deposition, Nutrient, Cycling and Soil Weathering on the Sustainability of Nine Forest Ecosystems." *Water, Air, and Soil Pollution*, 109(1–4): 101–135, 1999.

Wright, R.F., E. Lotse, and A. "RAIN Project: Results After 8 Years of Experimentally Reduced Acid Deposition to a Whole Catchment." *Canadian Journal of Fisheries and Aquatic Sciences*, 50(2): 258–268, 1993.

Visibility and Air Quality

Tropospheric pollution is often visible. Meteorologists and air quality specialists quantify pollution levels in terms of how well we can see. The clarity of the atmosphere is referred to as *visibility* and can be thought of as the farthest distance at which a person can see a landscape feature. It is often expressed as the visual range, or distance, at which a given feature can be seen and appreciated. Visibility is often expressed in terms of *deciviews*, a unit used to quantify perceived changes in visibility over the entire range. A deciview of 0 corresponds to clear air; deciviews greater than 0 represent proportional increases in visibility impairment. In general, a 1 deciview change would be noticed by most people regardless of initial visibility conditions. Visibility is closely related to air quality; degradations in visibility indicate increased gas or particle concentrations in the troposphere. These gases and particles are typically the result of human activities, such as combustion of fossil fuels in power plants and automobile engines.

Visibility degradation results when sunlight is scattered by gases and particles in the atmosphere. A portion of this scattering is natural. Our sky is blue, for instance, because gas molecules in the atmosphere are particularly effective at scattering the blue component of sunlight. Natural visibility conditions are considered to be 8 to 11 deciviews (about 100 to 120 kilometers or 60 to 80 miles) in the eastern United States, and 4.5 to 5 deciviews (about 175 to 185 kilometers or 110 to 115 miles) in the western part of the country. These numbers are approximate since monitoring efforts are relatively recent: visibility observations before the addition of anthropogenic pollutants to the atmosphere are largely unavailable.

In the eastern United States, carbon-based particles contribute to about 20 percent of the visibility impairment. Sulfates, another human-made pollutant, contribute to about 60 to 70 percent of this impairment. In the West, where there is a lesser reliance on coal combustion, sulfate and carbon-based particles may each contribute about 30 to 50 percent to the total visibility degradation. The third largest contributor to visibility impairment is typically wind-blown dust. In urban areas, nitrates can significantly increase the level of this degradation.

In the 1977 amendments to the Clean Air Act, the U.S. Congress established the goal of preventing future degradations in visibility and of remedying any existing visibility impairments at 156 national parks and wilderness areas. The National Park Service, U.S. Forest Service, Environmental Protection Agency (EPA), Fish and Wildlife Service, and Bureau of Land Management are working in coordination to achieve these goals. One part of this coordinated effort is IMPROVE (Interagency Monitoring of PROtected Visual Environments). The IMPROVE network provides instrumentation and protocols for monitoring visibility and air quality at several sites in the United States and around the rest of the world. In this way, data can be gathered concerning sources of impairment on local, regional, and national scales. IMPROVE began in 1987 with twenty long-term monitoring sites; today it includes over forty sites in parks and wilderness areas throughout the nation.

Results from IMPROVE and other monitoring efforts have shown clear links between sulfur emissions and visibility impairment on regional scales. Researchers at the National Park Service, U.S. Forest Service, and EPA have developed sophisticated mathematical techniques to identify and understand the emission sources that contribute to these impairments. The result has been the development of regional-scale air quality monitoring capabilities for evaluation of source-specific air quality impairments over scales of several hundred kilometers.

In some national parks, air quality is no better than in major cities, and regional haze is a primary contributor to this problem. Original EPA regulations were designed to limit air pollution traceable to a specific source. Combined pollution from a host of sources, including cars, power plants, and unpaved roads, was therefore not controlled. This led to a continued decline, rather than an improvement, in air quality in many of the parks. In 1990, Congress instructed EPA to produce measures that would ensure "reasonable progress" toward cleaning the air in parks. In 1997, the agency proposed regulations to approach the problems through a regional scale, rather than local scale, approach. On Earth Day 1999, Vice President Al Gore announced the Clinton administration's plans to participate in this process. The current plans entail phasing in stricter controls on power plants and automobile emissions, with the intent of restoring air quality in the parks by 2064.

Air quality is important not only from the aesthetic vantage of visibility, but also in terms of human health effects, most commonly related to respiratory problems. In 1997, almost 107 million Americans, accounting for almost 40 percent of the country's population, were reported to live in areas that failed air quality standards for at least one major pollutant. The Natural Resources Defense Council reports that each year, some 64,000 people may die prematurely from cardiopulmonary conditions linked to air pollution. Their analysis indicates that health complications due to poor air quality can shorten people's lives by an average of one to two years. Los Angeles tops the list with an estimated 5,873 early deaths per year, followed by New York (4,024 deaths), Chicago (3,479 deaths), Philadelphia (2,599), and Detroit (2,123). Air pollution problems and concerns are not confined to the United States. Episodes of extremely severe air pollution have occurred in Bangladesh, Iran, Indonesia, Hong Kong, and India in recent years. By U.S. standards, such conditions would be intolerable. In 1999 the World Resources Institute named Mexico City the world's most dangerous city for children. This rating comes from the city's poor air quality; one report indicates that the children there rarely use the color blue when painting a picture of the sky. In Santiago, Chile, children are often kept indoors to keep them from breathing the smoggy air.

Urban air quality problems are largely a result of chemical reactions among various pollutants from automobile emissions. Nitrogen oxides, carbon monoxide, and other compounds known as reactive hydrocarbons react in the presence of sunlight to create a brownish haze layer called photochemical smog. The primary ingredient of this smog is ozone. Ozone has beneficial effects when present in the stratosphere, but high levels in the troposphere can produce toxic results and has been linked to damage to plants. This "brown cloud" often veils the cities of Los Angeles, Mexico City, Santiago, and Denver. In these areas, topography, weather, and demographics are especially favorable for the formation of photochemical smog. The San Bernardino Mountains northeast of Los Angeles, for example, trap smoggy air over the inland valleys. In 1988, ozone concentrations in some of these valley cities, such as Pasadena, Azusa, and San Bernardino, exceeded clean air standards on more than 140 days. Since the early 1990s, conditions have been improving, but not without substantial expenditures to control pollutants and help limit exposure to hazardous air.

In the United States, the federal government sets standards for

Figure 1.10. Automobiles and smog choke a Mexico City street. Poor air quality can lead to health complications and even death in cities all around the world. © Stephanie Maze/CORBIS.

limiting air pollution and defining clean air. Individual states may also impose their own standards, which are often more stringent than those of the federal government. The standards set thresholds for levels of major pollutants including carbon monoxide, nitrogen dioxide, ozone, particulate matter, sulfur dioxide, and lead. The standards attempt to define the levels of these pollutants that might be considered relatively safe for an average person in good health. Responsibility for monitoring these pollutants is usually taken on by local or regional authorities, including various regional air quality districts. Scientists and officials at these agencies can also produce forecasts of smog levels and provide recommendations for lessening the severity of these episodes or limiting exposure. For instance, people may be restricted from using wood-burning sources of heat or be encouraged to car-pool to reduce emissions. In the worst smog areas, schools may be closed to protect children.

In 1997, the EPA conducted a review and reevaluation of the national ambient air quality standards. These reviews are required at least once every five years and involve a significant scientific and technical assessment of each pollutant and its environmental and health effects. As a result of the 1997 review, the EPA has established a new 8-hour ozone standard to help protect against exposure for prolonged periods. Standards for particulate matter are also being revised to set limits on particles smaller than 2.5 microns (about 10^{-4} inches) in diameter. These particles are small enough to be easily inhaled and therefore present significant threats to the human respiratory system. These problems are particularly acute for persons with emphysema or asthma. Current air quality standards, including the revisions for particulate matter, are available in print or on-line from the EPA and from regional air quality districts or councils and have also been published in a number of textbooks.

The EPA released its latest air quality trends report in April 2000 and reports that throughout most of the United States, air quality continues to improve. Observations indicate that from 1989 to 1998, lead concentrations decreased by 56 percent, carbon monoxide fell by 39 percent, and coarse particulate matter (including dust, dirt, and soot) fell by 25 percent. Nitrogen dioxide concentrations were reduced by 14 percent, and smog dropped by 4 percent. Despite these improvements nationally, air quality in many eastern locations was reported to worsen, especially in rural locations. Smog was observed to increase at seventeen of twenty-four National Park Service monitoring locations, and fine particle concentrations increased at seven of ten eastern rural monitoring sites.

The EPA is taking a number of steps to reduce smog in the United States. These steps include stringent rules to reduce car and truck tailpipe emissions, including new emission standards for all heavy-duty trucks and sports utility vehicles (SUVs), use of cleaner fuel, and setting standards for small hand-held engines. The EPA is also seeking to bring cleaner air to millions of Americans in the eastern United States by reducing the amount of pollution that is regionally transported, affecting air quality several states away from its origin.

References
Callahan, J.R. *Recent Advances and Issues in Environmental Sciences.* Phoenix: Oryx Press, 2000.
Environmental Protection Agency, <http://www.epa.gov>.
National Park Service, <http://www2.nature.nps.gov>.
National Resources Defense Council, <http://www.nrdc.org>.
Turco, R.P. *Earth Under Siege: From Air Pollution to Global Change.* Oxford: Oxford University Press, 1997.

CLIMATE

Weather conditions, air quality, and other aspects of atmospheric behavior all go toward determining a location's climate. *Climate* is defined as the average of weather over some location over a relatively long period of time (usually on the order of thirty years or more). Climate also includes statistics pertaining to extremes in weather over a similar period. Climate can most easily be thought of as how we expect our weather to behave. If you live in a part of the country where the average nighttime temperature in October is around 39 degrees, then it's a good bet that you will want to have your furnace serviced and operational by then—based on the normal climate for your location. Climate has received a lot of attention recently, and as a result, many of us are probably at least somewhat familiar with climate and climate change issues. In this section, we look at the science surrounding some of these issues and examine how knowledge of the working of earth's climate system can be tied to current events. We also look at scientists' responses to these issues and at new discoveries and understandings.

Urbanization

In regions of high population density and industrial activity—that so-called urban sprawl which is often all too familiar—the effects of human influences on climate are easily apparent and readily abundant. One of the most easily documented effects of urbanization is the tendency of cities to be warmer than the surrounding countryside. This "heat island" effect is most noticeable in the center of cities and results from the combination of unnatural physical characteristics and large energy consumption. Urban heating is most apparent at night, and during the summer months it can lead to increased use of air conditioners. Increased air conditioner use results in greater energy consumption, which increases the effect of the urban heat island—a so-called positive feedback.

Urbanization can transform natural landscape into a jungle of steel and concrete. The paved surfaces of city streets prevent water from being absorbed by soil and keep much of the remaining subsoil moisture from evaporating. In cases where local sewer systems lack sufficient runoff and storage systems, flooding after heavy rainstorms can be common. Tall buildings and other rigid structures alter the roughness of the land surface, causing a change in local wind patterns. Wind speeds in urban areas are often over 10 percent lower than unimpeded countryside winds. Decreased wind speeds contribute to the reduction of normal evaporation rates. They can also change natural temperature patterns and substantially alter the dispersion of smoke and other pollutants.

Urban climate modifications exhibit more than a local effect. Warm air is less dense than cool air. Thus, warm air over urban areas may rise upward at an above-normal rate, which can affect precipitation patterns over the city and for an unknown distance downwind. Statistical studies completed by Stanley Changnon and Floyd Huff at the Illinois State Water Survey reflect up to a 15 percent increase in precipitation in areas downwind of Chicago, Gary, Indiana, and Saint Louis, Missouri. This phenomenon is attributed to the increased particle matter emitted by car exhaust and industry. These particles serve as centers for the formation of droplets, which are eventually precipitated out.

References

Changnon, S. "Inadvertent Weather Modification in Urban Areas." *Bulletin of the American Meteorological Society*, 73(5): 619–627, 1992.

Huff, F.A., and S.A. Changnon, Jr. "Potential Urban Effects on Precipitation in the Winter and Transition Seasons at St. Louis, Missouri." *Journal of Climate and Applied Meteorology*, 25(12): 1887–1907, 1986.

Greenhouse Effect and Global Warming

While urbanization affects weather and climate on local and regional levels, there are other stimuli that can affect the atmosphere on a global scale. The idea of global warming has received widespread attention in recent years, although certainly the concept of climate change is nothing new. In 1816 Thomas Jefferson observed that "the climates of the several states of our union have undergone a sensible change since the dates of their first settlements." He also noted that since the time of Augustus Caesar, the climate of Italy had "changed regularly at the rate of one degree of Fahrenheit's thermometer for every century." Variations in climatological conditions for a given region are quite common and have often been documented in the historical literature. Recognition of human-induced climate change as a serious, modern environmental issue can be traced to work begun by David Keeling, then at Scripps Institute of Oceanography. In 1958, Keeling began making routine measurements of atmospheric carbon dioxide concentrations at the Mauna Loa Observatory in Hawaii. These measurements have become part of one of the most well-known graphical illustrations in climate research. Figure 1.11 shows carbon dioxide concentration in parts per million per volume as a function of time from 1958 to 1998. The curve illustrates the seasonal cycle due to photosynthesis and respiration of vegetation, as well as a distinct, long-term upward trend. Prior to 1950, the release of carbon dioxide to the atmosphere was mainly due to the oxidation of organic matter exposed by agricultural tilling. Now, the largest net source of atmospheric carbon dioxide is combustion of fossil fuels. This source adds an estimated 3 billion tons of carbon dioxide to the atmosphere yearly, the highest known increase of any substance in some 200 millennia. The clearing of forests for agriculture and other uses also contributes to the increased carbon dioxide loading, with the amount estimated to be 20 to 60 percent that of combustion.

Carbon dioxide, along with other trace gases such as methane and water vapor, is a greenhouse gas. These gases play an important role in the heating of earth due to their ability to absorb radiation. Radiation reaches earth in the form of solar energy from the sun. Part of this energy, about 20 to 25 percent, is reflected back to space by the clouds and atmosphere. Another 20 percent is absorbed by the atmosphere, allowing about half of the total incoming radiation to reach the earth's surface. Of this, approximately 18 percent is reflected back to the atmosphere and space, and the remainder is ab-

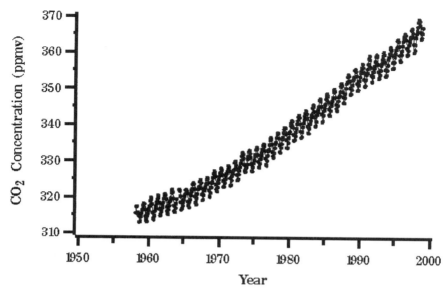

Figure 1.11. Atmospheric carbon dioxide concentrations measured at Mauna Loa, Hawaii, 1958–1998. The Mauna Loa record shows a 16.1 percent increase in the mean annual carbon dioxide concentration, from 315.83 parts per million by volume (ppmv) of dry air in 1959 to 366.70 ppmv in 1998. The 1997–1998 increase in the annual growth rate of 2.9 ppmv represents the largest single yearly jump since the Mauna Loa record began in 1958. Created by Dave Keeling and Tim Whorf, Scripps Institute of Oceanography.

sorbed. The absorption of solar radiation warms the earth's surface. Any object with heat will emit radiation, so radiation is in turn emitted to the atmosphere by the earth's warmed surface. The radiation is emitted primarily in the infrared part of the spectrum, where greenhouse gases are particularly absorbing. As carbon dioxide, methane, and other greenhouse gases absorb infrared radiation, they must also emit radiation. The added emission of radiation by gases in the atmosphere is termed the *greenhouse effect* and is the component responsible for making our planet a comfortable habitat for current life on earth. Without this effect, the earth's surface temperature would be approximately 30°C (60°F) degrees cooler, unable to support life as we know it.

Global warming refers to an expected increase in the earth's surface temperature as increasing accumulations of greenhouse gases (GHGs) in the atmosphere enhance the greenhouse effect. This change to our planet's climate would be largely human induced, as most of the increased GHG accumulation stems directly from fossil fuel combustion and other human activities. Global warming has been a topic of much attention and debate in recent years, with results of many different studies being proffered as evidence for and against climate changes. The most unified consensus is considered to be that of the Intergovernmental Panel on Climate Change (IPCC). Established in 1988 by the World Meteorological Organization, the IPCC is a multinational group of scientists, economists, and hundreds of other experts charged with examining the climate system and the social and economic impacts of human-induced changes. The panel released its second assessment report detailing evidence and effects of anthropogenic climate change and possible mitigation strategies in 1995. In general, the results indicate a 1.0 to 4.5°C (1.8 to 8.1°F) warming of the planet by 2100, a temperature warmer than any experienced so far in human history. The group's assessment also presents clear evidence for an observed 0.5°C (0.9°F) warming over the twentieth century.

In March 2000, the National Academy of Sciences in the United States released its consensus on apparent discrepancies in global temperature records. The disagreement had dated back decades, and people, whether supporting or disagreeing with the idea of human-induced global warming, were able to twist the results either way. The controversy had roots in differences between satellite data, which infer radiation-based temperature measurements, and direct measurements of temperature from buoys and other surface instruments.

In the 1980s, scientists John Christy at the University of Alabama and Roy Spencer at NASA's Marshall Space Flight Center in Huntsville began compiling a long-term temperature record from Microwave Sounding Unit instruments aboard earth-orbiting satellites. Analysis of this record suggested a cooling trend. The results were immediately presented as evidence that the threat of global warming had been overemphasized and that the warmth of 1997 may have been merely a random or anomalous event. Surface measurements, analyzed by James Hurrell and Kevin Trenberth, both of the National Center for Atmospheric Research, showed a warming trend of 0.12°C (0.22°F) per decade for the same period as the satellite record. Scientists had long known there were differences between the satellite and surface temperature measurements, but for the first time, this disagreement reached the greater scientific community, as well as the public, in a way that caught people's attention.

Scientists began lining up on both sides of the issue, supporting whichever set of results agreed with their own points of view. In an interview with Richard Kerr, warming skeptic Patrick Michaels at the University of Virginia called the satellite data "our only true global record of lower atmosphere temperature," claiming that the surface data must be flawed. Other scientists said that surface and satellite measurements could not be expected to provide identical values, since they do not even measure the same quantity. The observations used by the National Climatic Data Center and other agencies are recorded at various locations at the surface of the earth, while the satellite instruments measure microwave radiation emitted by oxygen molecules in the atmosphere and use these values to infer atmospheric temperature.

In June 1998, C. Prabhakara and colleagues at NASA's Goddard Space Flight Center published a new analysis of the satellite data. Their results indicate a warming of 0.12°C (0.22°F) per decade, within an error range of ±0.06°C (0.11°F). Prabhakara's group added corrections to the temperature record by accounting for errors resulting from the splicing together of measurements from the series of nine different Microwave Sounding Unit–carrying satellites. Two months later, Frank Wentz and Matthias Schabel at Remote Sensing Systems in Santa Rosa, California, published the results of orbital decay corrections they had added to the data set, doubling the uncertainty range in the results but showing a warming trend of 0.07°C (0.13°F) per decade. Christy and Spencer found that incorporation of these corrections in their analysis reduced the cooling trend to only −0.01°C (0.02°F) per decade. This is similar to an analysis by NOAA climatologist Jim Angell, who analyzed temperature measurements recorded by twice-daily weather balloon launches. For the years 1979 to 1996, the balloon measurements indicated a cooling trend of −0.02°C (0.04°F) per decade temperature trend. When Angell extended this analysis back to the balloon record's 1958 beginning, however, the results indicated a warming trend of 0.16°C (0.29°F) per decade, providing evidence of the record length's role in identification of trends.

Christy's latest calculations attempt to compensate for the relatively brief twenty-year period of the satellite record by removing the effects of ocean warming by El Niños and atmospheric cooling due to particulate matter or aerosol from volcanic eruptions. His new results show a warming of 0.03°C to 0.10°C (0.05 to 0.18°F) per decade, comparable to the trend reflected in the surface observations and supporting claims that our climate is indeed warming.

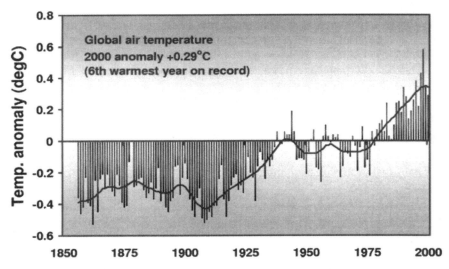

Figure 1.12. Global air temperature, 1850–2000. Data from P.D. Jones, M. New, D.E. Parker, S. Martin, and I.G. Rigor, "Surface Air Temperature and Its Variations over the last 150 Years," *Reviews of Geophysics* 37:173–99, 1999. Used by permission of Climatic Research Unit, University of East Anglia, 2001.

The compatibility of results of several independent temperature analyses may help establish their validity, but these results alone cannot answer all the questions. Do the reported warming trends necessarily prove the idea of an increased greenhouse warming caused by anthropogenic increases in atmospheric gases such as carbon dioxide and methane? The January 2000 report released by the National Academy of Sciences concludes that evidence of global warming cannot be ignored. The report indicates that the warming at the earth's surface is "undoubtedly real" and that in the past two decades, surface temperatures have risen at a rate that is substantially greater than the average for the past 140 years (see Figure 1.12). The panel points out that global warming processes can be masked by natural variability but indicates that the earth's temperature is nonetheless rising. Still, there were unresolved questions concerning how much of the increase was attributable to human activities and how much to natural causes. In the National Academy of Sciences report, John Wallace, professor of atmospheric sciences at the University of Washington and chair of the National Research Council Panel on Reconciling Temperature Observations, states, "The rapid increase in the Earth's surface temperature over the past 20 years is not necessarily representative of how the atmosphere is responding to long-term, human-induced changes, such as increasing amounts of carbon dioxide and other 'greenhouse' gases. The nations of the world should develop an improved climate monitoring system to resolve uncertainties in data and provide policy-makers with the best available information."

Despite trends in the temperature records that parallel the 0.10°C (0.18°F) per decade warming predicted by climate models, some scientists are cautious about viewing the surface warming as evidence of a strengthening greenhouse. The earth's climate has been known to exhibit large natural variability. Moreover, many aspects of climate and climate change, including the role of low-level pollutants and haze, are not well understood. Still, there are many scientists ready to point out that the bulk of the evidence does seem to suggest undeniably that global temperatures are increasing and that this increase can only be linked to an increase in anthropogenic greenhouse gases. Many research efforts are now focused on determining the effects of this warming on various regions of the globe.

Climate Models

Knowledge of how our climate may be changing comes from complicated models of how the atmosphere, ocean, and land surface behave and interact. These models rely on a large number of parameters that specify things like the chemical makeup of the atmosphere, the distribution of heat over the earth, and the roles of the oceans and land surfaces.

There are many models and types of models in use, and an overview is provided by Fred Singer in his book *Global Climate Change*. Models concerned with the processes of energy transport involved in climate, including the absorption of solar energy and emission of infrared, or heat, radiation to space, are referred to as energy-balance models (EBMs). These models deal with flows of energy and provide direct calculations of surface and atmospheric temperatures. Some EBMs, known as statistical-dynamical models, are able to parameterize atmospheric motions. Because these models require less computation time than general circulation models, they are very useful for studies over long timescales.

Radiative-convective models (RCMs) focus on the exchange of solar and infrared radiation at different levels of the atmosphere and among clouds, the atmosphere, and the surface of the earth. The models are extremely detailed in the calculations of radiative transfer processes and provide reliable estimates of temperature as a function of height.

General circulation models (GCMs) are the most comprehensive climate models and the type generally referred to in discussions of future climate change. These three-dimensional models express in mathematical terms the processes that dictate the general circulation, or motion, of the atmosphere and oceans. Given initial conditions, a GCM can simulate a subsequent time evolution predicting temperatures, winds, and precipitation as functions of space and time. GCMs are outgrowths of the numerical weather prediction models used in weather forecasting and are able to provide general results over timescales for which more detailed forecast calculations would be unreliable.

Heat, Drought, and Other Climate Indicators

In addition to models, information on how climate behaves also comes from direct observations. Surface weather data are collected routinely at many locations worldwide, and satellites provide a global picture of various quantities throughout the entire atmosphere. Recent data suggest that in many locations, and for the planet as a whole, surface temperatures are rising.

In the words of one Associated Press writer, the reports of each month of 1998 being the warmest on record were beginning to sound like a broken record. The question is, Are these reports true? And if they are, should they be considered the result and evidence of global warming? The excitement began in early 1998 when investigators at the National Climatic Data Center (NCDC) in North Carolina reported 1997 as the warmest of the century. Their analysis showed averaged combined land and ocean temperatures for 1997 to be 0.42°C (0.76°F) higher than the mean annual temperature for the years 1961 to 1990. Researchers at the British Meteorological Office and NASA's Goddard Institute for Space Studies in New York confirmed NCDC's findings and showed 1997's record-breaking temperatures to be part of a trend consistent over the past two decades. Several months in 1998 also set records as the warmest ever recorded, and global land and ocean temperatures for January through September averaged more than 0.69°C (1.25°F) above the 1880–1997 long-term normal.

In 1999, the global warming banner waved again, and in some

places was the only breeze to be felt. From July 19 until August 1, the eastern United States sizzled under temperatures approaching, and in many cases topping, 55°C (100°F). The excessive heat led to the deaths of at least 191 persons in twenty states. Rainfall was 50 to 80 percent below normal in several mid-Atlantic and northeastern states, causing drought conditions comparable to those experienced in Oklahoma in the 1930s. Maryland reported its driest one-year period since 1965–1966, with July 1998 to June 1999 rainfall a whole 30 centimeters (12 inches) short of the normal yearly amount. Water levels of all five Great Lakes were as much as 50 centimeters (17 inches) below 1998 depths, and subnormal rainfall was also reported in Alaska, Hawaii, Nevada, Idaho, Utah, and northern California.

Abnormal or anomalous climate conditions can show up in a variety of ways. On the night of June 22, 1999, meteorologist Richard Keen observed some unusual clouds visible over his home in Colorado. Composed of ice crystals located at an altitude of about 80 kilometers (50 miles), these noctilucent clouds appear as silvery-blue bands and shine brightly long after the sun has set. Observations by cloud researcher Mark Zeilcik in Edmonton, Alberta, and Mike Taylor, a physics professor in Logan, Utah, corroborated the sighting. The observations were remarkable due to the fact that noctilucent clouds have not previously been observed so close to the equator, at 40°N latitude. In 1994, University of Colorado researcher Gary Thomas predicted that middle and upper atmospheric cooling, thought to be associated with a global warming of the lower atmosphere, would make these clouds visible over the continental United States by the twenty-first century. Still, it is a commonly held perception among climate scientists that no single event can be used to prove global warming conclusively. In an article by Charlie Brennan, Thomas himself is hesitant to use the clouds as evidence of an increased greenhouse, stating, "It could be an indication that we're polluting our atmosphere with greenhouse gases, or it could be part of a natural cycle of long-term trends."

So, what of the idea of increased greenhouse warming caused by anthropogenic increases in atmospheric gases such as carbon dioxide and methane? Certainly it is a subject of much controversy. In general, although the trends in the temperature records parallel the 0.10°C (0.18°F) per decade warming predicted by climate models, some scientists are cautious to view this as evidence of a strengthening greenhouse. Many aspects of climate and climate change, including the role of low-level pollutants and haze, are still not well understood.

Other researchers solidly support the relationship between observed temperatures and greenhouse warming, however, and are focusing their efforts on determining the effects of this warming on various regions of the globe.

More Controversy

It is usually agreed that the global average temperature has increased by 0.5°C (0.9°F) over the past century, as reported by the IPCC. Not so easy to establish are the reasons for this increase. Many scientists believe that human activities, including the burning of fossil fuels in automobiles and power plants, contribute to at least part of this warming. Others argue that the warming is simply due to natural variability; historically, the earth's climate has always changed and will continue to change. In an interview with Berny Morson of the *Rocky Mountain News*, Roger Pielke, Sr., a professor of atmospheric science at Colorado State University, calls global warming "a misnomer" and points out that climate changes may not necessarily be exclusively in the warming direction.

In the March 1999 issue of the *Bulletin of the American Meteorological Society*, Dennis Bray and Hans von Storch of the Max Planck Institute for Meteorology in Hamburg, Germany, released a report summarizing the global warming views of approximately 400 scientists in Germany, the United States, and Canada. The report was a subset of a larger survey completed in 1996 that summarized the views of climate scientists in the United States, Canada, Germany, Italy, and Denmark. In general, scientists in the United States and Canada were most likely (by 62 percent and 70 percent, respectively) to agree with the statement that a global warming process is underway. While 59 percent of Canada scientists indicated agreement that climate change is mostly the result of anthropogenic causes, only 32 percent of U.S. scientists felt this statement had any degree of certainty. When asked about the degree to which climate change would have detrimental effects of some societies, scientists of all nationalities agreed: the percentages indicating a moderate to great degree of detrimental effects were on the order of 90 percent or greater. Interestingly, approximately 57 percent of Danish scientists and 33 percent of those surveyed from Canada indicated a belief that climate change would have a positive effect on the society in which they lived.

The indecisiveness concerning the climate effects of fossil fuel consumption and other human activities does not seem likely to be resolved soon. The climate is not a simple puzzle; as we will discuss in Chapter 4, many of the pieces still need to be better understood. Whatever the verdict on the global warming controversy, the National Oceanic and Atmospheric Administration reports that January–May 2000 set a record as being the warmest so far on record for the United States, with March through May being the warmest spring ever recorded. The Environmental Protection Agency has released reports of global warming impacts on health, coastal zones, agriculture, forests, water, and other natural resources. It also provides information on global warming effects on a state-by-state basis. The IPCC, a collaboration of 2,500 of the world's leading atmospheric scientists, released a 1998 report, *The Regional Impacts of Climate Change—An Assessment of Vulnerability*. These reports discuss the likelihood of sea level rise, shifting precipitation patterns, effects on crop yields, and other impacts of even a mild temperature increase.

References

Biagini, B. "Global Temperatures Rise to Record Levels in 1997." *Global Change*, 4(1): 8, 1998.

Bray, D., and H. von Storch. "Climate Science: An Empirical Example of Postnormal Science." *Bulletin of the American Meteorological Society*, 80(3): 439–455, 1999. Web site: Max Plank Institute for Meteorology, <http://w3g.gkss.de/G/Mitarbeiter/storch/thyssen//surveyintro.htm>.

Brennan, C. "Ice-Crystal Clouds Spotted for First Time Over State." *Denver Rocky Mountain News*, June 25, 1999, p. 7A.

Christy, J.R., and R.T. McNider. "Satellite Greenhouse Signal." *Nature*, 367: 325, 1994.

"Eastern U.S. Cools, But Not Much Rain." MSNBC Weather, Aug. 2, 1999.

Gelbspan, R. *The Heat Is On—The High Stakes Battle over Earth's Threatened Climate*. Reading, MA: Addison-Wesley, 1997.

Hougton, J. *Global Warming—The Complete Briefing* (2nd ed.). Cambridge: Cambridge University Press, 1997.

Hurrell, J.W., and K.E. Trenberth. "Difficulties in Obtaining Reliable Temperature Trends." *Journal of Climate* 11(5): 945–967, 1998.

Kerr, R.A. "Among Global Thermometers, Warming Still Wins Out." *Science*, 281(5385): 1948–1949, 1998.

MacIlwain, C. "Global-Warming Skeptics Left Out in the Cold." *Nature*, 403: 233, 2000.

Morson, B. "Global Warming Misnamed." *Denver Rocky Mountain News*, Jan. 9, 2000.

National Climatic Data Center. <http://www.ncdc.noaa.gov/ol/climate/research/1998/sep/sep98.html>.

National Research Council. *Reconciling Observations of Global Temperature Change*. Washington, DC: National Academy Press, 2000.

Prabhakara, C., et al. Global Warming Deduced from MSU. *Geophysical Research Letters*, 25(11): 1927–1930, 1998.

Singer, F. *Global Climate Change: Human and Natural Influences.* New York: Paragon House, 1989.

Spencer, R.W., and J.R. Christy. "Precise Monitoring of Global Temperature Trends from Satellites." *Science*, 247: 1558–1562, 1990.

"U.S. Drought Worsens." Jim Morris, CNN Weather, July 28, 1999.

Wentz, F.J., and M. Schabel. "Effects of Orbital Decay on Satellite-Derived Lower-Tropospheric Temperature Trends." *Nature*, 394: 661–664, 1998.

Chapter Two

Meteorological Technology

Many of the advances in meteorological understanding presented in Chapter 1 would not have been possible without significant progressions in technology. The twentieth century saw the development of radio, television, the computer, and the Internet. These tools alone have changed the face of modern communication and data exchange. Today's meteorologists work with a suite of new technologies for data collection, analysis, and dissemination. This chapter examines the development of computer technology in relation to weather and climate forecasting, provides an overview of many of the new instruments available for collecting data, and gives examples of how these new technologies are being used in scientific studies of the atmosphere and its phenomena.

COMPUTER TECHNOLOGY AND IMPROVEMENTS IN FORECASTING AND CLIMATE MODELING

The first computer able to perform the sophisticated mathematical calculations needed to solve equations describing motion of atmosphere was the ENIAC (Electronic Numerical Integrator and Computer), completed in December 1945 at the Ballistics Research Laboratory in Maryland. Its primary purpose was to compute trajectories of bombs and cannon shells during World War II. Shortly after ENIAC's introduction, the Electronic Computer Project (ECP) was begun at the Institute for Advanced Studies in Princeton, New Jersey, with the goal of producing a new and improved computer. This computer would do two important things. First, it would allow for software, essentially programs, to be stored within its memory. Second, it would be able to handle parallel processing, defined most simply as the ability to carry out multiple tasks at the same time.

Parallel processing is extremely essential for weather forecasting. Meteorologist Jule Charney, who joined ECP in 1948, had successfully run a computational model on ENIAC to generate a 24-hour forecast. The model unfortunately required 24 hours of computation time. On the new parallel processing machine, however, the forecast took just 5 minutes to produce. Charney's work led to the development of a relatively simple numerical weather prediction model that was released to the Weather Service for operational use in 1955. This model had been tested on about a dozen case studies, and its results were not necessarily accurate or reliable. But use of this model in an operational context did serve to unite modelers with practicing meteorologists, and this union led to the development of a suitable numerical forecast system as early as 1958.

From the 1950s onward, weather forecasts improved significantly as faster computers were able to handle more complicated physics and increased resolution, or grid spacing. In its first 33 years of operation, the National Meteorological Center acquired six state-of-the-art supercomputers, each about six times faster than its predecessor, allowing for an overall improvement of 10,000 times in speed as well as storage capacity. Improvements of similar magnitudes have not been seen in numerical forecasting, however. As discussed in Chapter 1, forecast accuracy is measured in terms of skill, basically a measure of how far off the forecast was in predicting the atmospheric pressure at a given height over the entire computational area of the model. Since the 1950s, forecast skill has improved from scores of approximately 50 to scores in the mid-20s. This skill score varies with the lead time of the forecast. For instance, National Meteorological Center models may be able to achieve a skill near 20 on a 36-hour forecast, while a 72-hour forecast may receive only a 50. Over the years, one solution proposed has been to increase the resolution of the models, though the benefits of such undertakings have been argued. In an article for *Weather and Forecasting*, Frederick G. Shuman, a former director of the National Weather Service, found that "doubling the resolution of a forecast only improves the skill of a forecast by about 15%, and even then only if the other characteristics of the model are suitably enhanced." Such a doubling of grid resolution must be accounted for in all three dimensions, therefore requiring an order of magnitude increase in the storage capability of the computer.

References
Fishman, J., and R. Kalish. *The Weather Revolution: Innovations and Imminent Breakthroughs in Accurate Forecasting.* New York: Plenum Press, 1994.
Kalnay, E. Numerical Weather Prediction. 1998. <http://weather.ou.edu/ekalnay/NWPChapter1/NWPChapt1.html>.
Shuman, F.G. "History of Numerical Weather Prediction at the National Meteorological Center." *Weather and Forecasting* 4(3): 286–296, 1989.

NEW TECHNOLOGIES AND CAPABILITIES

Many of the advances in meteorology would be impossible without the development of new technologies to explore and analyze the atmosphere. Accurate weather forecasting depends very strongly on the speed and memory of available computers. Forecast models require the best possible measurements of current conditions to serve as a basis for calculations into the future. The new technologies explored in this section offer a sampling of the resources that scientists can now use to understand and predict the behavior of the atmosphere.

New Computing Systems

One step toward increasingly powerful computing capabilities is evidenced by the new High Performance Computing System, acquired in spring 2000 by the National Oceanic and Atmospheric Administration's Forecast Systems Laboratory. The system is able to perform about 300 billion arithmetic operations per second, representing a twenty-fold improvement over the laboratory's previous computing system. Computing speeds are measured in terms of floating point operations per second (flops). With a final upgrade in 2002, the system will be able to handle as many as 4 trillion arithmetic computations—over four teraflops—per second. Laboratory director A.E. MacDonald says the new system will play a key role as researchers develop the next generation of weather prediction models. The computer will also benefit the North American Observing System (NAOS), a program designed to improve upper-air observations in the twenty-first century.

In August 1999, the Scientific Computing Division at the National Center for Atmospheric Research (NCAR) acquired a $6.2 million

Figure 2.1. The High Performance Computing System "JET," recently acquired by NOAA's Forecast Systems Laboratory. Courtesy of Wilfred von Dauster, National Oceanic and Atmospheric Administration (NOAA).

distributed-shared-memory computer. A distributed-shared-memory computer consists of many nodes, with memory shared by processors within the nodes. The supercomputer provides 2.5 terabytes (2.5 million bytes) of disk space and 1024 megabytes of memory, and can operate at speeds of 204 billion floating point operations per second. NCAR and other laboratories will share the computer, which is a first step toward a clustered computing network that will dramatically boost the center's computer power.

References
National Center for Atmospheric Research, <http://www.ucar.edu/ncar/>.
National Oceanic and Atmospheric Administration, <http://www.noaa.gov>.

Advanced Weather Interactive Processing System

Recent upgrades at the National Weather Service will allow forecast office personnel to take full advantage of the information these new computing capabilities can provide. The National Weather Service's $4.5 billion modernization, completed in 1999, included a new computer system providing direct links to 152 locations throughout the country. This Advanced Weather Interactive Processing System (AWIPS) replaces technology used since the 1970s. With AWIPS, meteorologists can combine weather radar, satellite imagery, hourly observed weather conditions, and numerical forecast output on a single computer screen rather than on multiple terminals. This ability to access and study multiple types of data efficiently is extremely important to the National Weather Service's forecasting process, especially when related to hurricanes, tornadoes, or other severe weather.

The advanced information processing, display, and telecommunication capabilities of AWIPS provide the first opportunities to integrate all meteorological, hydrological, satellite, and radar data fully and easily. These capabilities allow forecasters to prepare and issue more timely and accurate forecasts and weather and flood warnings.

The new technology offered by AWIPS has received widespread recognition. In December 1997, *Popular Science* named AWIPS one of 100 Best of What's New award winners. In June 1999, AWIPS received top honors in the Computerworld Smithsonian Award program's Environment, Energy, and Agriculture category.

References
National Oceanic and Atmospheric Administration Forecast Systems Laboratory, <http://www.fsl.noaa.gov>.
National Weather Service, <http://www.nws.noaa.gov>.

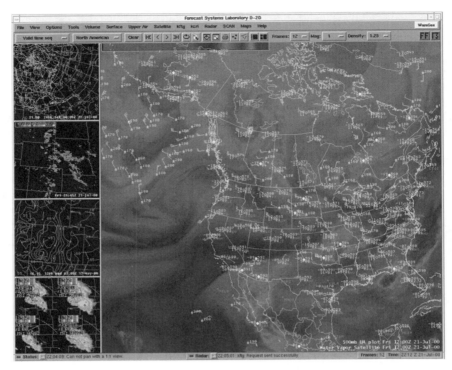

Figure 2.2. Screen shot from the Advanced Weather Interactive Processing System (AWIPS), a new computer system used by the National Weather Service to analyze up-to-the-minute weather information. Courtesy of the National Oceanic and Atmospheric Administration (NOAA).

Next Generation Radar

The National Weather Service modernization also included installation of Weather Surveillance Radar (WSR-88D) systems known as NEXRAD, or Next Generation Radar. The development of weather radar has been perhaps one of the most exciting advances in ground-based observing systems. By identifying severe weather phenomena in their initial stages, weather *radar* (*ra*dio *d*etecting *a*nd *r*anging) systems have been credited with saving hundreds and even thousands of lives each year worldwide.

Weather radar systems produce images of precipitation in the atmosphere by measuring the radar signals reflected by raindrops, hailstones, snowflakes, and other forms of condensed water vapor. The WSR-88D relies on the Doppler effect, a physical phenomenon marked by a shift in the signal's frequency depending on whether the object under observation is moving toward or away from the observer. The images provided by Doppler radar give forecasters a detailed look at a storm's behavior by displaying rainfall or precipitation intensities, along with information on the speed and direction of winds. These radars have been in use for weather applications since the mid-1960s. They are especially useful for identifying storms and locations within storms that may produce tornadoes. The WSR-88D system used by the National Weather Service helped improve the average lead times for tornado warnings from approximately 5 minutes in 1986 to almost 12 minutes by 1998.

References
Beach, R. "Doppler Weather Radar: Benefiting from Innovation." *World Meteorological Organization Bulletin* 47(2): 132–138, 1998.
National Weather Service, <http://www.nws.noaa.gov>.

Lidar

Lidar, an acronym for *li*ght *d*etection *a*nd *r*anging, is the optical analog of radar. Lidar systems work by transmitting high-frequency laser pulses through the earth's atmosphere. When a laser pulse comes into contact with a gas molecule, aerosol particle, cloud droplet, or other scattering object, a portion of the pulse will be scattered back to the instrument. By measuring the time this process takes, along with the amount of power returned, scientists can learn a great deal about the object doing the scattering and its position in the atmosphere.

Several meteorological applications of lidar systems are discussed in this chapter. Wind measurement is one such application and can be performed very well using a Doppler lidar instrument. Doppler lidar systems work similarly to Doppler radar systems (see the WSR-88D above), but measure changes in frequency of laser light rather than changes in sound waves. The National Oceanic and Atmospheric Administration's Environmental Technology Laboratory operates a powerful ground-based Doppler lidar system able to estimate wind speeds at distances of over 24 kilometers (15 miles). This lidar system was deployed during the Mesoscale Alpine Programme (MAP), occurring between October 1 and November 15, 1999, in Innsbruck, Austria. The measurements obtained by the lidar system are used to improve understanding of the strong winds blowing through gaps in mountain ranges, mountain passes, and other low topographic points.

References
Liou, K-N. *An Introduction to Atmospheric Radiation.* San Diego, CA: Academic Press, 1980.

National Oceanic and Atmospheric Administration Environmental Technology Laboratory, <http://www.etl.noaa.gov>.

Wind Profilers

In addition to the WSR-88D, other types of radar systems are useful in a variety of meteorological applications. In the late 1980s, two prototype wind profiler units were built and installed, one near Bloomfield, Connecticut, and the other near Platteville, Colorado. A wind profiler is an upward-looking Doppler radar able to measure how wind speed and direction change with altitude. Two research laboratories of the National Oceanic and Atmospheric Administration (NOAA), the Aeronomy Laboratory and the Wave Propagation Laboratory (now the Environmental Technology Laboratory), performed the first implementations of the wind profiler instruments. The sensitivity of these instruments was so impressive that over the next 18 to 20 months, thirty more units followed. The network of instruments was known as the Wind Profiler Demonstration Network (WDPN), which in the mid-1990s became the NOAA Profiler Network (NPN).

As of December 1999, the NOAA Profiler Network, managed and operated by NOAA's Forecast Systems Laboratory, involved thirty-two instruments located in key locations across the United States. In addition to providing information on the vertical structure of wind, the NPN instruments use radar and acoustics to obtain data on the vertical structure of temperature and other quantities. Weather forecasters can use these data to improve their predictions. For instance, the Weather Service Forecast Office in Boulder, Colorado, uses the profiler data to forecast the beginning and ending of rain and snow events in mountainous areas. This information can be useful in many areas of the West and to highway maintenance agencies throughout the country. The NPN data can also help provide improved and timely forecasts of severe storms or high wind events.

References

National Oceanic and Atmospheric Administration Forecast Systems Laboratory, <http://www.fsl.noaa.gov>.

National Weather Service and Office of Oceanic and Atmospheric Research. *Wind Profiler Assessment Report and Recommendations for Future Use*, 1987–1994. Silver Spring, MD: Author, 1994.

Figure 2.3. One of the thirty-two wind profilers operational in the United States as of 1999. These upward-looking Doppler radar systems provide valuable information on wind speed and direction throughout the lower atmosphere. Courtesy of the National Oceanic and Atmospheric Administration (NOAA).

Automated Surface Observing Systems

The Automated Surface Observing Systems (ASOS) program is a joint effort of the National Weather Service (NWS), the Federal Aviation Administration (FAA), and the Department of Defense (DOD). The ASOS program serves as the primary surface weather observing network in the United States, providing data on cloud coverage, visibility, precipitation, wind, temperature, pressure, and other meteorological variables. The main emphasis of ASOS is to support weather forecast activities and aviation operations, but the data also have many applications in meteorological, hydrological, and climatological research.

ASOS works nonstop, updating observations every minute, 24 hours a day, every day of the year. The information can help the NWS increase the accuracy and timeliness of its forecasts and warnings and also provides critical weather data to improve aviation safety. ASOS routinely and automatically provides computer-generated voice observations directly to aircraft in the vicinity of airports, using the Federal Aviation Administration's ground-to-air radio. These messages are also available via a telephone dial-in port.

Reference
National Oceanic and Atmospheric Administration Forecast Systems Laboratory, <http://www.fsl.noaa.gov>.

Aircraft Communications Addressing and Reporting System

The Aircraft Communications Addressing and Reporting System (ACARS) provides automated weather reports from commercial aircraft. These data provide forecasters with observations over areas where conventional meteorological observations may not be available. The data can also provide vertical profiles of temperature, wind direction, and wind speed at altitude intervals of only a few hundred feet. These vertical profiles are critical for many meteorological applications, including monitoring changes in the freezing level during winter storms and monitoring the strength and depth of winds aloft.

Airlines contributing ACARS data include American, Delta, Federal Express, Northwest, United, and United Parcel Service. Although most of the observations are collected during the daytime and over the continental United States, the participation of parcel-carrying airlines ensures some nighttime coverage as well. More than 500 aircraft send some 50,000 observations to NOAA's Forecast Systems Laboratory each day. Scientists decode and quality control the data for use as input to the Rapid Update Cycle (RUC) and Eta forecasting models. The observations also provide sufficient coverage to help initialize global-scale models.

References
Martin, G. *Examples of the Advantages of ACARS Data*. Western Regional Technical Attachment No. 00–07, Apr. 11, 2000.
National Oceanic and Atmospheric Administration Forecast Systems Laboratory ACARS, FAQ page, <http//acweb.fsl.noaa.gov/FAQ.html>.

Atmospheric Radiation Measurements

Solar radiation reaching the earth and thermal radiation emitted by earth can be measured by *radiometers*, instruments that convert radiant energy to a form of energy that can be more easily quantified. This form may include thermal or electrical energy. The Surface Radiation Budget Network (SURFRAD) of the National Oceanic and Atmospheric Administration's Air Resources Laboratory currently operates six radiation measurement sites throughout the United States. The network was established in 1993 with the goal of supporting climate research through accurate, continuous, and long-term radiation monitoring. Instruments at the sites obtain measurements of incoming solar and infrared radiation levels, as well as data on the amount of radiation that is reflected or emitted by the earth's surface. The instruments are able to distinguish between direct and scattered solar radiation and also provide measurements of potentially harmful ultraviolet (UV) radiation levels.

The SURFRAD observations are complemented by measurements of incoming visible and UV radiation at nine ISIS (Integrated Surface Irradiance Study) stations also operated by the Air Resources Laboratory. Other national networks include the Cooperative Network for Renewable Resource Measurements (CONFRRM). Operated by the National Renewable Energy Laboratory in Golden, Colorado, CONFRRM measures solar radiation and wind measurements at thirteen sites in the United States. The U.S. Department of Energy maintains Cloud and Radiation Testbed (CART) sites in northern Oklahoma, in the tropical western Pacific, and on the northern slope of Alaska. These sites are part of the Department of Energy's largest global change research program and are focused on obtaining field measurements and developing models to understand solar and thermal infrared radiative processes in the atmosphere and at the earth's surface.

Because ultraviolet (UV) radiation levels at the earth's surface may be directly affected by losses of UV-blocking ozone in the stratosphere, accurate measurements of these amounts are an area of special concern. Exposure to UV radiation can have detrimental effects for humans and other organisms, as is discussed in greater detail in Chapter 3. To address this need and provide data for research on UV effects, UV radiation measurements are being obtained at fourteen national parks and in seven urban areas as part of an interagency agreement between the U.S. Environmental Protection Agency and National Park Service. The data from these twenty-one sites can provide valuable information of changing UV levels and ecosystem responses. The UVB Monitoring Network operated by U.S. Department of Agriculture provides additional data to address effects of UV on crops for human and animal consumption. The Department of Agriculture's monitoring activities are conducted at six research sites as well as a climatology network proposed to consist of thirty to forty sites nationwide.

References
Environmental Protection Agency UV Network, <http//www.epa.gov/uvnet/>.
Integrated Surface Irradiance Study, <http://zephyr.atdd.noaa.gov/isis/isis.htm>
National Oceanic and Atmospheric Administration Surface Radiation Research Branch, <http://www.srrb.noaa.gov>.
National Renewable Energy Laboratory CONFRRM Network, <http://rredc.nrel.gov/solar/new_data/confrrm>.
U.S. Department of Agriculture UV Network, <http://uvb.nrel.colostate.edu/>.
U.S. Department of Energy CART sites, <http://www.arm.gov/docs/sites.html>.

Cloud Detection Instruments

An important element to any understanding of radiation, and therefore climate, is the role played by clouds. Ground-based radar and lidar systems can be extremely useful for obtaining cloud observations. In late 1996, researchers at the Cloud and Radiation Testbed (CART) site in northern Oklahoma began operating a new instrument for measuring clouds. Designed by the National Oceanic and Atmospheric Administration's Environmental Technology Laboratory and the Pacific Northwest National Laboratory, the millimeter-wave cloud radar (MMCR) is able to provide long-term, detailed observations of cloud structure and evolution. The MMCR relies on the same type of signal processor as used by wind profilers. The observations will greatly improve quantitative records of cloud conditions over the CART site and help researchers better understand how clouds affect climate. The MMCR-type radar systems also have applications for aviation weather, weather modification, and basic cloud physics research.

The Depolarization and Backscatter Unattended Lidar (DABUL) instrument works similarly to the MMCR-type systems, using light waves instead of sound. The DABUL instrument can provide information on cloud geometry and on the phase (gas, liquid, or ice) of water in the cloud. It can also provide information on the size and amount of dust, soot, or other particles that may be present in the atmosphere. The DABUL instrument has been used in several field experiments, including the Surface Heat Energy Budget of the Arctic campaign, discussed below.

Total Sky Imager

In 1999, newly developed Total Sky Imager instruments, specially designed for viewing the sky to determine cloudiness, were installed at four Surface Radiation Budget Network (SURFRAD) stations. The Total Sky Imager (TSI) provides forecasters and the general atmospheric research community an easy-to-use and reliable system for continuous sky imaging. The system uses state-of-the-art camera optics and newly developed software to estimate the percentage of cloud cover in a 180-degree total hemispheric field of view. It can also compute the fractions of individual types of clouds, including thin cirrus, low-level stratus, or convective clouds.

The TSI represents a significant development in automated, ground-based cloud observations and can provide important data for radiation and climate research. The instrument provides quantitative estimates of cloudiness, replacing observations traditionally reported by trained human observers. As the number of trained observers decreases due to budgetary constraints, the objective estimates provided by the TSI have become increasingly important in forecasting and other applications.

References
National Oceanic and Atmospheric Administration Environmental Technology Laboratory, <http://www.etl.noaa.gov>.
National Oceanic and Atmospheric Administration Surface Radiation Research Branch, <http://www.srrb.noaa.gov>.
Yankee Environmental Systems, <http://www.aero-laser.de/yes4.html>.

Atmospheric Gas Measurement Instruments

Atmospheric gases, including concentrations of carbon dioxide, ozone, and ozone-depleting substances, can be measured using a variety of techniques. One of these techniques is differential absorption lidar (DIAL). The DIAL approach transmits two laser wavelengths through the atmosphere and examines the difference in absorption of the two wavelengths. The difference is due to the presence of molecules that absorb radiation more effectively at one wavelength than at the other, providing information on both the type and the amount of gas present. DIAL techniques can be used to measure amounts of water or ozone in the atmosphere.

Atmospheric gases can also be measured using in situ (meaning "in the original place") techniques. The most common in situ method for atmospheric gas measurements involves using flasks to capture a sample of air at a particular altitude. The air sample can be analyzed in the laboratory to determine the types of molecules present. Measurements of this type are useful in determining changes in the amounts of atmospheric carbon dioxide or of ozone-depleting gases.

Atmospheric ozone is a gas that has been a topic of much concern, especially because of the severe depletion observed in polar regions. Understanding trends in ozone depends on accurate measurements of ozone concentrations. Ozone observations began with some isolated measurements in the 1920s, and systematic observations have been conducted since the 1940s. The instruments used for these observations were Dobson ozone spectrophotometers, which are still the backbone of the Global Ozone Observing System. The ozone spectrophotometer, perfected by George Dobson, calculates the amount of absorption at the wavelengths of light most sensitive to ozone. The Dobson total ozone records are one of the longest geophysical measurement time series currently in existence.

Routine ground-based ozone measurements are made from almost 100 observatories and are now complemented by global observations from satellites. Specialized ozone-measuring instruments flown in space include the Total Ozone Mapping Spectrometer (TOMS), the Stratospheric Aerosol and Gas Experiment (SAGE), and the Solar Backscatter Ultra Violet spectrometer (SBUV).

Several newer techniques can also be used to obtain ozone measurements. One laser instrument, the Balloon-borne Laser In Situ Sensor (BLISS), can fly aboard a balloon high into the stratosphere to provide information on ozone and other stratospheric trace gases. Ozone measurements can also be obtained by UV photometers. These instruments, similar to the Dobson instrument, measure absorption of incoming radiation and are usually operated from the ground. However, some photometers are small enough to fly aboard research aircraft or helium-filled balloons to provide in situ ozone observations.

The ozonesonde is another type of instrument commonly used to obtain in situ observations. Ozonesondes are relatively small, measuring only about 4 inches by 6 inches, and can be carried by balloons high into the stratosphere. The Climate Monitoring and Diagnostics Laboratory of the National Oceanic and Atmospheric Administration launches these instruments once a week from locations at the South Pole, the Galapagos Islands, Fiji, Hilo, Samoa, and in the United States in California, Colorado, and Alabama. The sonde instruments compute ozone amount based on an electrochemical reaction between ozone and iodine. In the Antarctic, ozonesondes are carried aloft by temperature-resistant plastic balloons that expand as the surrounding air pressure drops. The balloons eventually reach an enormous size: about 70 to 80 feet high and about 60 feet in diameter.

Tropospheric ozone, the type that air quality officials are concerned with, is measured with a UV photometric ozone analyzer. The ozone analyzer has been the standard instrument used by the U.S. Environmental Protection Agency since the 1970s and works by measuring the change in absorption of light beam internal to the instrument.

References
Johnson, B. Personal communication, 2000.
National Aeronautics and Space Administration, <http://www.nasa.gov>.
National Oceanic and Atmospheric Administration Climate Monitoring and Diagnostics Laboratory, <http://www.cmdl.noaa.gov>.

ADVANCES IN SATELLITE TECHNOLOGY

In 1959 and 1960, two successes in meteorological instrumentation and satellite technology provided a global view of earth unmatched by any other observing system available at that time. The first, the launch of Explorer 7 in October 1959, provided brand-new data on solar radiation reflected by the earth and infrared radiation emitted by the earth. The instrument responsible for these measurements, a radiometer (radiation measuring instrument) developed by Verner Suomi and colleagues at the University of Wisconsin, was the first of its kind to orbit the earth successfully. In 1960, the return of the first satellite views of earth and its weather systems by TIROS 1 (Television and Infrared Observational Satellite) marked the beginnings of a new era in meteorological observation. In the forty years since, satellite technology has gained acceptance as an indispensable tool for forecasting and climate studies. Between 1990 and 1994 alone, approximately fifty new meteorological satellites were launched by such countries as France, Japan, the United States, the former Soviet

Figure 2.4. Atmospheric reserachers launch an ozonesonde instrument to study the ozone hole above Antarctica. The instrument is carried by balloon upward through the atmosphere to heights of more than 40 kilometers. Courtesy of the National Oceanic and Atmospheric Administration (NOAA).

Union, India, Korea, and Sweden. The suite of instruments onboard today's satellites can measure vertical profiles of temperature, ozone and other atmospheric gases and can provide global data on the distribution of clouds, aerosols, and other atmospheric constituents.

Satellite meteorological instruments rely on a variety of *remote sensing* techniques, meaning that the measurements are made from afar or remotely, as opposed to within or in situ to the quantity being measured. Remote sensing uses measurements and interactions of radiation at various wavelengths in the electromagnetic spectrum. Measurements involving no interaction with the radiation itself, for example, use of the Suomi radiometer, are referred to as *passive sensing* techniques. The images returned by TIROS 1, which were basically camera images of visible radiation, are also examples of a passive remote sensing technique. Radar systems—in which a signal is actually transmitted, interacts with the medium being measured, and is received by the instrument—are an example of an *active sensing* technique. These systems, along with microwave instruments and optical radars that have also been used in space, are sensitive enough to provide extremely detailed and accurate measurements but have a higher power consumption than passive instruments. Many of the newest observational satellites employ combinations of passive and active instruments aboard single platforms. Using this method, investigators can obtain the most thorough information possible while minimizing costs associated with satellite launch and operation.

In the 1980s, NASA began plans for a program called Mission to Planet Earth. The program's objective involved the examination of air, water, land and biota on a global scale in an attempt to understand how natural processes affect us, the inhabitants of the earth, and how we in turn affect these natural processes. This program has grown into an international collaboration involving science and technical contributions from the United States, Canada, Europe, Japan, and Russia. In January 1998, Mission to Planet Earth was renamed the Earth Science Enterprise. Phase I of the Earth Science Enterprise included focused, free-flying satellite projects, space shuttle missions, and various airborne and ground-based studies. Phase II began in 1999 with the launch of the first Earth Observing System (EOS) satellite, *Terra*, originally named AM-1.

The meteorological objectives of *Terra* include studies of cloud and aerosol properties, radiative energy fluxes, tropospheric chemistry, and atmospheric temperature and humidity. In accordance with the goal of obtaining long-term global observations, *Terra* will also provide data on land cover and land use, fires, volcanoes, land and ocean surface temperatures, sea ice, and snow cover.

The *Terra* platform is home to five instruments. The Advanced Spacebourne Thermal Emission and Reflection Radiometer (ASTER), provided by the Japanese Ministry of International Trade and Industry, will obtain high-resolution images of land surface, water, ice, and clouds using fourteen spectral channels from visible to thermal infrared. The Clouds and Earth's Radiant Energy System (CERES) instrument is a broadband radiometer measuring contributions of the earth's atmosphere, clouds, and surface to the overall radiation budget. The Multi-angle Imaging Spectroradiometer (MISR) and Moderate Resolution Spectroradiometer (MODIS) each uses multiple channels to obtain data on top-of-atmosphere, cloud, aerosol, land surface, and vegetation properties. The fifth sensor, Measurements of Pollution in the Troposphere (MOPITT), performs measurements of carbon monoxide and methane in the lower atmosphere. The combination of these instruments aboard the *Terra* platform marks the beginning of a new era of interdisciplinary and comprehensive monitoring.

Another Earth Science Enterprise mission seeks to study the interaction of water vapor, clouds, precipitation and the earth's climate system. An estimated two-thirds of the world's total precipitation falls in the tropics, and the release of latent heat associated with this rainfall accounts for three-quarters of the energy that drives the global atmospheric circulation. Until the Tropical Rainfall Measurement Mission (TRMM), precipitation and precipitation rates in the tropics and subtropics were extremely difficult to measure, with uncertainties often as high as 50 percent. The scarcity of land-based measurement sites and large ocean areas in these regions prevent ground-based observation of rainfall amounts as they occur. By providing a three-dimensional view of precipitation and energy exchange in the tropical and subtropical latitudes from 35°S to 35°N, TRMM seeks to fill these gaps.

TRMM consists of a suite of five instruments, each expected to remain in operation for at least three years. One of these is the Precipitation Radar (PR). This active sensing system is the first rain radar ever to fly in space and works by transmitting a microwave signal at approximately 13.8 gigahertz through the atmosphere and clouds to the earth's surface below. Part of this signal is reflected back to the receiver onboard the satellite. The weakening or attenuation of the signal strength provides information on the presence of cloud droplets or raindrops along its path. Using this information, scientists can

Figure 2.5. NASA's *Terra* spacecraft comprises five state-of-the-art sets of instruments to collect data for continuous, long-term records of the earth's land, oceans, and atmosphere. The instruments operate by measuring sunlight reflected by the earth and heat emitted by the earth. Courtesy of the National Aeronautics and Space Administration (NASA).

obtain a detailed view of rainfall structure and more accurately estimate rainfall amounts over land or ocean. The high-resolution data obtained by the PR will provide unique information of the three-dimensional structure of tropical rainfall and will complement measurements from two other sensors: the TRMM Microwave Imager and the Visible Infrared Scanner.

The TRMM Microwave Imager (TMI) is a five-channel passive microwave radiometer. Heavy rainfall is often associated with very deep, vertically developed clouds whose tops may contain many large ice particles. These ice particles tend to scatter the microwave radiation emitted naturally by the earth's surface and by smaller water droplets within clouds. Scientists use observations at the five microwave channels to deduce cloud precipitation consistent with measurements at each frequency. TMI provides the primary precipitation measurement data for the mission. The third primary instrument onboard the satellite, the Visible Infrared Scanner (VIRS), is a five-channel scanning radiometer. VIRS measures in the visible and infrared parts of the spectrum, with wavelengths from 0.63 to 12 microns. It provides information on vertical profiles of condensation heat release associated with raindrop formation. Cloud radiation measurements are also among the data obtained by VIRS. As of September 1998, scientists had reported agreement to within 25 percent for total rainfall from these three primary instruments.

Also aboard the TRMM platform is a CERES instrument comparable to that aboard EOS AM-1 and the Lightning Imaging Sensor (LIS). LIS is an optical telescope and imaging system whose purpose is acquisition of information on cloud-to-cloud and cloud-to-ground lightning occurrences. This information will be used to investigate the correlation of lightning with rainfall and other storm properties.

Other new technology to gain orbit will be the spaceborne radar launched aboard NASA's Cloudsat in 2003. This radar will slice through clouds to provide important information on their vertical structure. Cloudsat's cloud profiling radar will be the first space-borne instrument of its kind and will also provide high-quality observations of clouds on a global basis. The mission is an international collaboration of the United States, Canada, Germany, and Japan. The data will complement observations from the EOS-PM satellite. To be flown in conjunction with Cloudsat is PICASSO-CENA, a cooperative mission between the United States and France. PICASSO-CENA will study the effects of thin, transparent clouds and aerosols on solar-energy transfer. PICASSO-CENA will use a spaceborne lidar to complement the EOS-PM measurements and provide a unique three-year global data set of cloud and aerosol properties.

The National Oceanic and Atmospheric Administration has historically operated several satellites and recently launched NOAA-15 in March 1999. NOAA-15 represents the latest in a series of polar-orbiting satellites used to make twice-daily weather and climate observations for all points on the globe. It is also armed with a new channel that will help scientists distinguish clouds from snow. Although not a difficult task for our eyes on earth, differentiating clouds and fog from snow and ice in satellite observations made 800 kilometers (500 miles) above the earth can be a significant challenge. The 1.6-micron measurement channel aboard NOAA-15 has helped make this process much easier. Methods for using the 1.6-micron channel in conjunction with two other measurement channels to discriminate more accurately among snow, ice, clouds, and fog in the satellite imagery were recently developed by Rob Fennimore, George Stephens, and coworkers at NOAA's National Environment Satellite, Data, and Information Service. Launch of a similar satellite in the 1999–2000 winter marked the beginning of the 1.6-micron channel's routine operational use.

During the next twenty years of satellite observations, corresponding to about the time period for NASA's Earth Science Enterprise, instrument improvements and new developments are inevitable. The role and importance of satellite technology in global monitoring, especially from a climate perspective, cannot be argued. Many years have passed since that long-ago day when the first TIROS 1 images showed our earth from above, yet the excitement of satellite meteorology remains full of promise.

References

Earth Observing System (EOS) home page, <http://eospso.gsfc.nasa.gov>.
EOS AM-1 home page, <http://eos-am.gsfc.nasa.gov>.
Kidder, S.Q., and T.H. Vonder Haar. *Satellite Meteorology: An Introduction*. Orlando, FL: Academic Press, 1995.
Kummerow, C., et al. "The Tropical Rainfall Measuring Mission (TRMM) Sensor Package." *Journal of Atmospheric and Oceanic Technology*, 15(3): 809–817, 1998.
Mission to Planet Earth and Earth Science Enterprise pages, <http://www.hq.nasa.gov/office/mtpe, http://www.earth.noaa.gov/missions/index.html>.
TRMM home page, <http://trmm.gsfc.nasa.gov>.

FIELD CAMPAIGNS

New technologies enable meteorologists to undertake intensive field studies on the atmospheric environment. The VORTEX (Verification of the Origin of Rotation in Tornadoes Experiment) project discussed in Chapter 1 is an example of one of these field programs. The research campaigns described below are typically international collaborations designed to answer numerous scientific questions. The campaigns combine observations from a diverse suite of instruments, including ground-based and in situ measurement systems, as well as satellites.

SHEBA

On October 2, 1997, the Surface Heat Budget of the Arctic Ocean (SHEBA) field experiment was begun from a drifting sea ice camp in the Arctic Ocean. To assess the interactions among sea ice, clouds, and atmosphere in the Arctic, the Canadian ice-breaker *Des Groseilliers* was intentionally frozen into an ice pack and set adrift. Data collection from the SHEBA ice station continued for one year while the ship and ice pack drifted over 4300 kilometers (2700 miles). The ice station scientists worked in teams of twenty to thirty people to gather data on many aspects of the Arctic environment. SHEBA scientists worked with a suite of instruments, including radiosondes to measure profiles of temperature, pressure, humidity, wind speeds up to 10 kilometers (6 miles) into the atmosphere; radar and lidar systems for assessing the distributions and physical properties of clouds; and radiometers, on the ground and aboard aircraft, to provide estimates of the amount of radiation reaching the Arctic surface and being reflected or emitted from the surface to the atmosphere. The scientists also collected data on sea ice thickness and on the temperature and composition of the ocean waters. These data, combined with observations from satellite, provide a wealth of information on the Arctic climate.

SHEBA, directed by Richard Moritz of the University of Washington, involves contributions from more than 150 investigators from the United States, Russia, Norway, and Canada. The project was initiated to develop, test, and implement models simulating Arctic ocean-atmosphere-ice processes and to improve the interpretation of satellite remote sensing data in the Arctic. The interactions among sea ice, surface properties, and clouds and radiation in the Arctic have historically been puzzling, although it has long been clear that these interactions exert a strong influence on the global climate. Satellite observations over the Arctic region are difficult to interpret due to problems in distinguishing cloud from cold snow and ice surfaces. With the data from SHEBA, scientists hope to be better able to simulate Arctic processes in climate models, which may lead to better predictions of potential global impacts of Arctic climate change.

Changes in the Arctic environment were reported prior to the SHEBA mission, and the earliest SHEBA results confirmed these reports. Scientists aboard the ice station noticed a thinning or complete absence of ice in many areas and also found that the depth of freshwater from melting ice was three times greater than the freshwater depths measured 22 years ago. Studies prior to SHEBA indicated changes in the variations of the ocean's temperature and salinity with depth, and changes in ocean currents and air motions of the region. SHEBA was spurred in part by these studies, in an effort to collect more detailed ocean surface and atmospheric observations to complement the earlier data. Analysis of these observations continues.

References
Goss Levi, B. "Adrift on the Ice Pack, Researchers Explore Changes in the Arctic Environment." *Physics Today*, 51(11): 17–19, 1998.
SHEBA, <http://sheba.apl.washington.edu>.

Indian Ocean Experiment

A major gap in the science of climate change prediction involves the role of small particles in the atmosphere, called *aerosols*, in contributing to regional and global change. Aerosols can be natural or human based in origin and are typically very small: their diameters are on the order of a millionth of a centimeter (10^{-7} inch) or less. Because of their ability to reflect and scatter sunlight back to space, these tiny particles can have noticeable regional cooling effects. Aerosols can also serve as building sites for clouds by providing surfaces on which water vapor can condense to form water droplets. The presence of aerosols can increase the number of water droplets in the cloud, which in turn makes for smaller cloud droplets. Changes in these microphysical cloud properties mean changes in the cloud's ability to reflect or absorb solar or terrestrial radiation, and these changes can then affect climate. The effects of aerosols on climate remain a complex issue and present one of the largest sources of uncertainty in

Figure 2.6. Instruments aboard the Indian R/V *Sagar Kanya* during the Indian Ocean Experiment (INDOEX). The instruments were used to measure aerosols, water vapor, solar radiation, and wind over the ocean near the Indian subcontinent. The rigid-hull inflatable boat from the *Ronald H. Brown* is tied up alongside the *Sagar Kanya*. Photo by Lieutenant Mark Boland, NOAA Corps. Courtesy of the National Oceanic and Atmospheric Administration (NOAA).

future climate predictions. To address this issue, atmospheric scientists from the United States, Europe, and countries adjoining the Indian Ocean began an intensive field campaign in the Indian Ocean region. The campaign was called the *Ind*ian *O*cean *Ex*periment (IN-DOEX).

INDOEX was begun in the late 1990s to collect data on aerosol composition, transport, and effects over the tropical Indian Ocean and Arabian Sea. The site chosen presented a unique natural laboratory for the study, delineating an area where clean air masses from the southern Indian Ocean met aerosol-laden air from the Indian subcontinent. In this area, an intensive field campaign was conducted from January 1999 to April 1999. INDOEX scientists used instruments aboard multiple ships and aircraft and on island stations to obtain measurements of aerosols and reactive atmospheric gases, solar radiation, wind, and water vapor distribution. These data can be used in conjunction with satellite observations to develop regional maps of aerosol effects.

Scientists from universities and national laboratories worldwide participated in INDOEX. Analysis of data collected during the intensive field phase of INDOEX is in progress as of this writing, but preliminary results confirm influences of human-based, or anthropogenic, substances in the Arabian and Indian Ocean regions. These substances, including dust and aerosols, ozone, carbon monoxide, carbon dioxide, and nitrogen oxides, are transported from the Indian subcontinent and can be detected in elevated quantities as far as 1500 kilometers (930 miles) offshore. One anthropogenic substance, non–sea salt sulfate, was found to increase by a factor of four. These changes in the level of pollutants can influence scattering and absorption of sunlight and thermal radiation in the atmosphere and have discernible impacts on regional climate.

Reference
Indoex: Indian Ocean Experiment, Center for Clouds, Chemistry, and Climate, Scripps Institute of Oceanography, University of California, San Diego, <http://www-indoex.ucsd.edu>.

Third Convection and Moisture Experiment

For seven weeks in the summer and fall of 1998, scientists from the National Aeronautics and Space Administration, the National Oceanic and Atmospheric Administration, and several universities participated in an experiment to study storms in the Atlantic Ocean. The Third Convection and Moisture Experiment (CAMEX-3) was designed to improve understanding of storm formation and behavior. Robbie Hood at NASA's Marshall Space Flight Center served as lead mission scientist for the experiment, which provided information on hurricanes Bonnie, Danielle, Earl, and Georges.

CAMEX-3 involved observations from the Tropical Rainfall Measurement Mission (TRMM). Specially designed laser instruments for use aboard DC-8 aircraft were also important in data collection. The observations brought some surprises. For instance, investigators found that the wind patterns flowing into and out of the upper altitudes of hurricanes were much more complicated than expected. The CAMEX-3 results are valuable largely because of their usefulness in developing better predictions of storm strength and duration. These predictions could potentially save lives and would also help to reduce hurricane evacuation zones along coastal areas.

Reference
National Aeronautics and Space Administration Marshall Space Flight Center, <http://ghrc.msfc.nasa.gov/camex3/mission_desc.html>.

Aerosol Characterization Experiment in Asia

ACE-ASIA (Aerosol Characterization Experiment in Asia) is a proposed monitoring program designed to integrate direct (in situ) observations, satellite measurements, and model calculations. The goal of ACE-ASIA is to investigate the roles of atmospheric particles called aerosols in climate and in the earth's biological, geological, and chemical cycles. The Asian-Pacific region is an area of significant coal and biomass burning. These anthropogenic activities have greatly perturbed atmospheric particle loading in this region, making it an appropriate site for measurements and monitoring. Data collected by ACE-ASIA researchers will help to reduce the overall uncertainty associated with calculations of aerosol effects on climate and will help to achieve sufficient understanding of the atmospheric chemical system. This understanding can be used for predictions of future aerosol-radiation interactions and climate response.

Reference

National Oceanic and Atmospheric Administation, Pacific Marine Environmental Laboratory, <http://saga.pmel.noaa.gov/aceasia/>.

SOLVE/THESEO 2000

Other recent or ongoing field experiments include SOLVE (the SAGE III Ozone Loss and Validation Experiment). The SAGE (Stratospheric Gas and Aerosol Experiment) III satellite was launched in 1999 to provide observations of ozone and aerosols in the global atmosphere. SOLVE was scheduled for four months during the winter of 1999–2000 to obtain correlative measurements of ozone, clouds, water vapor, particulates, and chemical species in the Arctic high latitudes. In close collaboration with SOLVE was THESEO 2000, the Third European Stratospheric Experiment on Ozone. Over 200 researchers from the United States, Canada, Europe, Russia, and Japan participated in the experiments. SOLVE and THESEO 2000 together represent the largest and most comprehensive field experiment ever to occur in the Arctic region. The campaigns used DC-8 aircraft to fly several 8- to 10-hour missions, each covering about 6400 kilometers (4000 miles), and collected valuable information on processes that control ozone loss in the Arctic and midlatitudes. The data also help scientists to quantify the rate of chemically induced ozone loss during the Arctic winter and spring.

References

SOLVE, National Aeronautics and Space Administration, <http://cloud1.arc.nasa.gov/solve/>.

THESEO Norwegian Institute for Air Research, <http://www.nilu.no/projects/theseo2000/> and European Ozone Research Coordinating Unit <http://www.ozone-sec.ch.cam.ac.uk>.

Tropospheric Ozone Production About the Spring Equinox

Another field campaign, addressing Tropospheric Ozone Production about the Spring Equinox (TOPSE), took place over mid- and high-latitude North America during spring 2000. TOPSE's goal was to investigate the chemical and dynamical evolution of chemical composition of the lower atmosphere, with particular emphasis on the springtime ozone maximum in the troposphere. The campaign consisted of a series of aircraft flights starting near Denver, Colorado, and reaching as far north as 80° latitude. The in situ and remotely sensed measurements of particulate matter and chemical species in the atmosphere will help scientists understand the potential feedback between chemistry and climate and the implications for global change.

Reference

University Center for Atmospheric Research Atmospheric Chemistry Division, <http://topse.acd.ucar.edu/>.

Cooperative Atmosphere-Surface Exchange Study–1999

The Cooperative Atmosphere-Surface Exchange Study–1999 (CASES–99) was set up east of Wichita, Kansas, in October 1999 to study the nighttime stable boundary layer. The boundary layer separates the free atmosphere from the surface. The nighttime boundary layers occur over a cold surface so are typically stable. CASES–99 experimenters deployed a High-Resolution Doppler Lidar (HRDL) to take measurements of these nighttime boundary layers and study their transition during the morning and evening. The data acquired will provide important information on processes within the stable boundary layer that are currently not well understood or were previously assumed to be negligible. Improved understanding of the exchange of heat, momentum, and atmospheric trace species through this layer is critical in accurately accounting for these processes in numerical models.

Reference

Cooperative Atmosphere-Surface Exchange Study–1999, Colorado Research Associates, <http://www.colorado-research.com/cases/CASES-99.html>.

FIRE III

As part of the International Satellite Cloud Climatology Project (ISCCP), the First ISCCP Regional Experiment (FIRE) began conducting research on cloud systems in 1984. In 1995, FIRE Phase III was initiated to conduct an Arctic cloud experiment over the Beaufort Sea off the north coast of Alaska. Called the FIRE III Boundary Layer/Arctic Cloud Component, the field campaign data will be used to assess the roles of Arctic cirrus and stratus clouds on the high-latitude climate system. The second part of FIRE III, the upper Tropospheric Cloud Component, studied large-scale cirrus anvils associated with storm development in the warm equatorial regions.

Reference

National Aeronautics and Space Administration, Marshall Space Flight Center, <http://wwwghcc.msfc.nasa.gov/ampr/fire3.html>.

Detect Icing with Polarization Experiment

The Detect Icing with Polarization Experiment (DIPOLE) is funded by the Federal Aviation Administration to apply advanced radar techniques to detection of hazardous icing conditions. Aircraft icing is blamed for an average of forty fatalities per year and is therefore an area of much concern to scientists and pilots alike. DIPOLE uses dual-polarization radar measurements to detect differences between water droplets and ice crystals. The DIPOLE research provides important information on the use of airport-based radar systems to detect ice and therefore issue warnings to pilots. Preliminary results also show that low-power radar systems can be made light enough to fly onboard an aircraft and detect icing conditions in route.

Reference

National Oceanic and Atmospheric Adminstration, Environmental Technology Laboratory, <http://www6.etl.noaa.gov/projects/dipole.html>.

OTHER MONITORING EFFORTS

In addition to information from field experiments, meteorologists depend on data from ongoing, often large-scale monitoring programs.

These efforts, of varying size and focus, are at various stages of implementation in many locations around the world. They usually involve multiagency or even multinational cooperation. We examine two of the monitoring programs proposed for ensuring continuity in atmospheric measurements.

North American Observing System

The North American Observing System (NAOS) is a multiagency effort geared toward guiding recommendations for future atmospheric observations. Participants include the U.S. Department of Defense, the Federal Aviation Administration, the National Aeronautics and Space Administration, the National Science Foundation, and the National Oceanic and Atmospheric Administration. NAOS came into existence to address the need for a systematic examination of observing capabilities on and above the North American continent. The initial focus of the program will be on upper-air observing systems, including the North American radiosonde network, meteorological satellites, weather surveillance radars, and automated reports from commercial jet aircraft.

Reference
National Oceanic and Atmospheric Administration Forecast Systems Laboratory, <http://www.fsl.noaa.gov>.

Global Air-Ocean In-Situ System

The Global Air-ocean IN-situ System (GAINS) is an ambitious program aimed at providing global, operational observations for use in weather prediction. The proposed program consists of a network of high-tech balloons evenly distributed above the earth at altitudes of 18 to 27 kilometers (60,000 to 90,000 feet). The operational program, proposed to begin in 2006, would involve a network of 400 balloons distributed over virtually every 10-degree square of latitude and longitude on the globe.

GAINS is being developed by the National Oceanic and Atmospheric Administration's Forecast Systems Laboratory, New Mexico State University's Physical Science Laboratory, and Global Solutions for Science and Learning. Test flights of early versions of these balloons indicate that they can be steered by changing altitude to locate areas of different wind shear. This ability to steer the balloons allows

Figure 2.7. A Global Atmosphere-ocean IN-situ Systems (GAINS) balloon launch. GAINS is a proposed observing system that will obtain atmospheric measurements from high-altitude balloons distributed over the earth's surface. Courtesy of Cecilia Girz, National Oceanic and Atmospheric Administration (NOAA).

them to be routed to areas of interest, for instance, near a volcanic eruption or over an oceanic algae bloom.

The balloons, each over 30 meters (100 feet) in diameter, would be equipped with measurement packages that would be dropped through the atmosphere to obtain data on a number of atmospheric and oceanic parameters. The measurements obtained could provide twice-daily observations of atmospheric temperature and moisture profiles, as well as data on carbon dioxide, ocean, and subsurface ocean parameters. These global observations would be extremely valuable for use in weather predictions, especially in increasing the accuracy of three-day and longer forecasts.

Reference
National Oceanic and Atmospheric Administration Forecast Systems Laboratory, <http://www.fsl.noaa.gov>.

TECHNOLOGY: AN AREA OF ONGOING ADVANCEMENT

The rapid advances in meteorological technology and increasing international collaboration on field experiments and observing campaigns help improve and expand scientists' understanding of the atmosphere. New computer systems and measurement techniques have been developed to help scientists understand an atmosphere so immense and so complex that it cannot be simulated in a laboratory. The field campaigns discussed represent only a sample of the measurement efforts that are underway. Researchers are always developing new instrumentation to perform new types of measurements. Experimenters will use these new measurements to understand another component of the large body of fluid that is our atmosphere. The results increase our knowledge of atmospheric dynamics, chemistry, radiation, and their interactions and can lead to improved predictions, from the local to global level, from time frames of several minutes to several decades.

Chapter Three

Meteorology and Society

Meteorological issues are important to all aspects of society. Global warming, ozone depletion, weather forecasting, and other issues discussed in Chapter 1 are topics of concern or debate among scientists, policymakers, and the public at large. Persons in the scientific community must collect the evidence, complete the research, and present reports on these issues. Is our planet warming? That is a question for the scientists. What are the effects if it is or does? That is a question of direct relevance to society, and a number of social scientists, economists, biologists, and other experts are working to find the answers.

How do the issues currently at the forefront of meteorological and atmospheric research affect society? This chapter provides information of potential use in answering this question. It begins with an examination of issues related to weather and weather hazards and continues with a discussion of potential impacts attributed to climate variability and climate change. The last section addresses human beings' attempts to modify the weather. Whether we control the weather or the weather controls us, the societal impacts, either beneficial or detrimental, can be extensive.

Many current issues in atmospheric science involve at least some measure of uncertainty, and this uncertainty complicates the understanding of societal implications. How can we know the effects of global warming on society when the only thing scientists agree on absolutely is that climate does vary? How do we anticipate the results of living under a depleted ozone layer when we do not know whether the next decade will bring a continuation of the depletion or the beginning of an ozone recovery? Meteorologists and atmospheric scientists continually seek an improved understanding of the thin layer of air that surrounds our planet. But not all questions currently have answers, and not every weather forecast is perfect. How much is society investing in efforts to achieve this understanding, and is the investment paying off? What are costs of weather and climate impacts on society? Who decides? As we shall see, sometimes the answers are not absolute.

WEATHER AND SOCIETY

The weather forecasts and warnings produced by meteorologists have intrinsic value to society. This value can be demonstrated in terms of prevention or reduction of losses of lives or property. Over the years, research organizations, insurance companies, and other interested agencies have copiously documented the costs of severe weather, including hurricanes, tornadoes, flash floods, droughts, blizzards, hail, and damaging winds that can result in massive devastation. Hurricane Hugo, which came ashore South Carolina in September 1989, destroyed more board feet of lumber than the Mount St. Helens eruption of 1980 and the Yellowstone National Park fires of 1988 combined. And Hugo was not even the costliest hurricane. That record is held by Hurricane Andrew, which caused more than $20 billion in damage in Florida and also caused damage in coastal Louisiana. Andrew left twenty-three people dead and over 250,000 homeless. But the death toll could have been much worse. A hurricane striking Galveston, Texas, in 1900 resulted in a death toll of more than 8,000. In 1998, Hurricane Mitch caused 11,000 deaths in Central America. Andrew's death toll was reduced largely because the National Weather Service was able to track the storm for days. Accurate and early warnings of its strength and movement helped people take life-saving precautions.

Each U.S. citizen, probably without realizing it, spends about $4 per year to support the National Weather Service. In exchange, the National Weather Service produces some 10 million forecast products each year, including more than 734,000 weather forecasts and 850,000 river and flood forecasts. It also issues between 45,000 and 50,000 severe weather warnings each year. These warnings are broadcast on NOAA Weather Radio and on television news stations, and can be instrumental in helping people to prepare for weather disasters.

Most importantly, they provide people in the at-risk areas an opportunity to get themselves and their families to safety.

In the United States, the number of fatalities directly attributable to weather has decreased over the past several decades. The total ranges from about 300 to 400 to over 10,000 weather-related fatalities each year, depending on how the numbers are counted. Single events, such as the heat wave of 1995, can increase this total by as much as 1,000. Improvements in forecasting help to decrease these fatalities: in the United States, the number of deaths from tornadoes decreased from 1,945 in the 1930s to 485 between 1986 and 1995.

Because of its size, geography, and varying terrain, the United States experiences more frequent severe weather than any other nation. Still, severe weather can be experienced in any part of the world. Between 1960 and 1989, weather disasters including drought and famine accounted for over 25 percent of all fatalities worldwide and affected over 92 percent of persons touched by any kind of disaster, including civil strife.

Damage from severe or extreme weather can be extensive and costly. The National Weather Service reports that between 1991 and 1995, an estimated $88.9 billion was spent on weather-related damages in the United States alone, with an annual average of $17.8 billion. This includes damage from hurricanes, tornadoes, floods, drought, dust storms, mud slides, avalanches, snow storms, heavy rain, strong winds, hail, lightning, extreme temperatures, and fire weather. There has been some speculation that these weather events have become more frequent or more intense in recent years. But the growing U.S. population and distribution of this population across the country may be a significant component in fostering this appearance. Recent years have seen increased numbers of people living in areas that tend to be most greatly affected. The number of U.S. residents living in coastal areas is continually increasing. These areas are particularly vulnerable to hurricanes, flooding, winds, and other severe weather.

Severe weather can have direct and indirect costs. When a hurricane comes ashore a populated section of coastline, for example, the direct costs of the damage can be fairly accurately assessed through insurance claims and other investigations. But what about production losses potentially incurred as businesses prepared for or cleaned up after the storm? What are the costs of a major evacuation? What if the evacuation was not needed? What if evacuation was needed but had not been ordered?

Figure 3.1. Traffic backs up along a Florida highway as residents evacuate to escape the heavy rains and high winds expected from Hurricane Floyd. © Reuters Newmedia Inc./CORBIS.

The National Oceanic and Atmospheric Administration reports that from 1980 through 1999, the United States sustained forty-four weather disasters, each causing $1 billion or more of damage. The vast majority (thirty-eight) of these disasters occurred from 1988 through 1999, with seven events happening in 1998 alone. Despite these numbers, the United States does not have a monopoly on severe weather. In fact, three of the twentieth century's four deadliest weather disasters occurred not in the United States but in China.

The list of the twentieth century's worst weather disasters, compiled by the National Oceanic and Atmospheric Administration, is topped by three disasters that heavily affected China, including a severe drought-induced famine, which claimed more than 24 million lives in 1907, and other drought-related famines in 1928–1930, 1936, and 1941–1942. The 1931 Yangtze River flood, ranked as the fourth worst weather disaster of the century, was responsible for some 3.7 million deaths. The list also includes drought in regions ranging from the Ukraine and Volga in the former Soviet Union, to India, to the Sahel in Africa.

In December 1999, mudslides devastated much of the Vargas region in Venezuela. Responsible for an estimated 25,000 deaths, the flooding and mudslides are expected to rank as one of the twentieth century's top ten weather catastrophes. The disaster began when torrential rains triggered floods and mudslides that crashed down Venezuela's Mount Avila, wiping out hillsides and valleys, and taking with them the communities newly established to accommodate the area's growing population. Caracas, Venezuela, was nearly devastated, with 150,000 to 200,000 people left homeless and thousands of bodies buried in the mud or washed out to sea. The hardest hit areas were covered by a layer of mud and debris up to 7 meters (22 feet) deep. The town of Carabelleda, once a plush resort area, lay buried under a wash of dirt, concrete, rocks, and tree trunks. Officials estimate it may take at least ten years to rebuild many of the devastated areas.

The floods and mudslides resulted from the heaviest rainfalls to hit Caracas and the northern states of Venezuela in 100 years. Officials at the United Nations Disaster Office in Caracas report that a significant amount of information was available about the dangers of flooding in many of the most damaged areas, but development in these areas was allowed to continue despite this knowledge. The dangers were increased by clear-cutting of trees for construction material and fuel and by excavating hillsides to allow construction, even in areas where such practices were prohibited by planning laws. In addition to the outstanding loss of life, the people of Venezuela are now faced with economic costs that far outweigh any short-term gains associated with the construction boom. The suburbs of Caracas have lost hundreds of millions of dollars in housing and business buildings and continue to lose tens of millions of dollars each day businesses remain out of operation.

Even with our modern infrastructure, and in some cases because of it, our societies can be extremely vulnerable to the effects of severe weather. The close of 1999 was marked by strong storms causing extensive damage in France. In the aftermath, more than eighty people had been killed, and some 3.4 million homes were without electricity. Winds approaching 124 miles per hour toppled tens of thousands of trees, injuring scores of citizens. The winds blew oil from a massive tanker spill off France's coast onto the western beaches. In southwestern France, two of four nuclear reactors were inundated by flood waters, and some 90,000 households were without running water. The same storms were responsible for over thirty deaths in Britain, Spain, Italy, and Switzerland and were attributed to damage in Germany and Belgium. U.S. Risk Management Solutions estimated that the storms could be Europe's costliest weather disaster, topping the $6.1 billion of damage from European windstorms in 1990.

As in Venezuela, the storms that pummeled Western Europe may result in serious economic consequences. The United Nations Economic Commission for Europe anticipates that the swath of uprooted trees stretching through France, Belgium, Switzerland, and Germany could be devastating to the timber industry; it could be years before the markets return to stability. Authorities in Switzerland report that the storms resulted in a loss of about 9.97 million cubic meters (10.9 million cubic yards) of timber. This figure is double the number of trees felled during the severe windstorms of 1990 and sets a new record, representing two years of normal timber harvest. In addition, because the wind-fallen trees are often unsuitable for construction or furniture uses, the market for recycled fiber faces being suddenly overwhelmed by a surplus of supplies. The fallen timber poses ecosystem dangers by enhancing the risks of insect infestation, which may spread to standing trees and further harm the region's forests. Infection among bird and animal populations once inhabiting the forests' trees could also lead to disastrous effects for the region's wildlife.

Floods

Historically, few weather-related phenomena can match the devastation and losses attributed to flooding. Floods can affect countries worldwide and in general are among the most costly of all weather-related phenomena. The Floodplain Management Group reports that parts or all of over 20,000 communities in the United States are at substantial risk of flooding. This flooding can occur along large rivers, along small streams, in the desert, and on hillsides. Under the right set of circumstances, almost any area is capable of being flooded.

In the United States, a growing population has led to more construction in flood-prone areas, which has stirred much debate over flood assistance programs and government flood management. From 1990 to 1997, floods contributed to more than $37 billion in property damage in the United States and led to the loss of 244 lives. Federal flood relief payments during this time were estimated at $15 billion to $30 billion. The National Flood Insurance Program, providing insurance to people in flood-risk areas, was $917 million in debt by the end of 1997. To many flood experts, such payments are

Figure 3.2. Volunteers look for weak spots in a dike protecting a Moorhead, Minnesota, home during the Red River floods of 1997. From 1990 to 1997, floods caused more than $37 billion in property damage in the United States alone. Courtesy of Dave Saville, Federal Emergency Management Agency (FEMA).

subsidies that allow people to occupy high-risk areas. When space was plentiful, people chose to build on high ground, out of harm's way. But population expansion has resulted in new subdivisions that stretch across low-lying areas or sit dangerously close to rivers and creeks.

Flooding is most often the result of heavy rain, but can also be caused by rapidly melting snow or by ice jams blocking the flow of rivers. The Red River flood of 1997 is a prime example of this type of springtime flooding situation. The Red River flows north along the North Dakota–Minnesota border into the southern part of Manitoba, Canada. The winter of 1996–1997 brought record amounts of snowfall to the Red River basin, an area whose terrain contains little change in elevation. In April 1997, the most severe blizzard in over fifty years dumped an additional 50 centimeters (20 inches) of snow on the region. As the snow began to melt, flooding began in southern portion of the basin and proceeded north, damaging 500 buildings in Breckenridge, Minnesota, threatening the flood-control efforts of Fargo, North Dakota, and eventually reaching the towns of Grand Forks, North Dakota, and East Grand Forks, Minnesota. On April 18, the dikes protecting the cities began to give way, inundating about half of Grand Forks and all of East Grand Forks. A fire gutted eleven buildings in downtown Grand Forks; in the end, damage to the two cities was estimated at over $2 billion. The flooding continued northward into Canada, although nowhere was the devastation as tremendous at that affecting the communities of Grand Forks and East Grand Forks.

The record-breaking flood levels on the Red River in 1997 and resulting damages focused new attention on flood issues in the United States. These issues concerned the role of flood predictions and the use and misuse of forecasts. The North Central River Forecast Center (NCRFC), operated in conjunction with the National Weather Service Forecast Office in Chanhassen, Minnesota, had issued its predictions for the Red River basin flooding up to two months in advance. But the numbers provided in mid-February were river heights expected based on average temperature and no precipitation and on average temperature and average precipitation. Roger Pielke, Jr., of the Environmental and Societal Impacts Group at the National Center for Atmospheric Research served as an independent member of the Survey Assessment Team that visited the area following the flood. Pielke's interviews with various decision makers in the river basin revealed that many officials interpreted the National

Weather Service predictions incorrectly and that the confusion was added to by a lack of estimates of the range of uncertainty on the forecast numbers.

These problems in interpreting the forecast river heights led to a misuse of the predictions by the public as well. Following 1979, when the Red River had reached the highest levels seen so far in the century, the city of Grand Forks began raising dikes and developing detailed flood-fighting plans. Thus, as reported by Pielke, although 95 percent of the residents were aware of flood insurance, 79.6 percent reported that the forecasts, while calling for unprecedented flooding, were not so alarming as to make them feel it would be necessary. On April 22, the river crested at about 16.5 meters (officially, 54.11 feet) in East Grand Forks, about 1.5 meters (5 feet) higher than predicted, and the town was under water.

The 1997 Red River flood was a disaster with important lessons for weather services providers, policymakers, and the public. The destruction and the events that followed provide information on the potential for devastation associated with floods and demonstrate the necessity of accurate, and clearly communicated, forecast information. With greater numbers of people now living in flood-prone areas, this information has become more essential than at any time prior in our history.

Federal Emergency Management Agency statistics indicate that floods account for up to three-quarters of all federal disaster declarations each year. In the 1990s alone, flooding affected forty-five of the fifty states and caused record property damage. Since 1993, major flooding episodes have occurred in the Mississippi River basin, in Texas, in California, in the Pacific Northwest, in North Dakota and Minnesota, and along the Ohio River. From 1993 to 1996, damages due to flooding totaled more than $28 billion and resulted in the loss of more than 300 lives. As shown by the statistics in Chapter 7, these numbers are greater than those associated with any other weather-related disasters occurring in the same time frame.

Droughts

While floods result largely from too much precipitation, droughts are hydrologic imbalances resulting from periods of abnormally dry weather. Like the too-wet conditions related to floods, droughts can be responsible for a far-reaching range of social, economic, and environmental impacts.

Perhaps the most extensive drought in U.S. history, the Dust Bowl of the 1930s dried up 202,000 square kilometers (125,000 square miles) of land from 1934 to 1941, forcing the relocation of thousands of farmers. Several years of below-normal rainfall led to a severe water shortage in the Great Plains in the 1950s, and in 1970, dry conditions led to drought in California, resulting in tens of millions of dollars of losses due to wildfires. In 1988, rainfall totals over the Midwest, northern plains, and Rocky Mountains were 50 to 85 percent below normal. Fires destroyed approximately 16,600 square kilometers (10,000 square miles) of forests in the Northwest, and 8,500 square kilometers (5,300 square miles) burned in Yellowstone National Park alone. Crops and farm animals died as farmlands turned to deserts, and by summer's end, the worst drought to occur in the United States in fifty years had affected thirty-five states.

The social and economic impacts of drought are manifested most broadly through public safety concerns and reductions in quality of life. The National Drought Mitigation Center reports that drought is one of the most important natural triggers of malnutrition and famine. Food shortages can lead to loss of life, as evidenced by the millions of famine-related deaths associated with drought episodes in China during the twentieth century. Drought conditions can also force the migration of a large portion of the population. Northeastern Brazil, a region often hard hit by drought, experienced a net loss of nearly 5.5 million people from 1950 to 1980. Much of this population shift was the result of migration from rural to urban areas, and in many cases, drought and the accompanying food shortages were primary factors in the decision to relocate. Population migrations of this sort can stress the social infrastructure of the urban area affected by the immigration, and can deprive rural areas of human resources necessary for economic development.

Even in countries with very strong economic systems, drought can result in losses in crop, dairy, and livestock production, leading to a myriad of other social and economic effects. Income losses for farmers can result in bankruptcy, which means fewer farmers to contribute to the food supply in future years. These income losses also place strain on financial institutions and can lead to federal, state, and local revenue losses as a result of a reduced tax base. Drought can affect navigability of streams, rivers, and canals and have a negative impact on fisheries production. In combination, the detrimental effects of drought can retard national economic growth.

Although drought is rarely a direct cause of death, it can lead to mental and physical stress, especially for those affected economically. Hot, dry weather can also bring an increase in respiratory ailments and other diseases. The elderly are often most susceptible to these health concerns. Dust and smoke in the air can reduce visibility, and range fires and wildfires can threaten homes and livelihoods. The 1988 drought led to an estimated $39–40 billion in economic losses, though experts believe that the estimate is too low and does not include the effects on human health or the disruption to farmers and rural communities.

Each year, some region of the United States experiences a severe or extreme drought. Unlike warnings for flooding or other types of severe weather events, warning times for drought conditions can be up to a year. However, often there is no warning that climate factors will conspire to create the abnormally dry conditions, and in some cases, an area can be experiencing a drought before the actual conditions are recognized. Grazing and agriculture, both affected by drought, can also contribute to the arid conditions if precautions are not taken to protect drought-sensitive areas. Identifying a drought onset is complicated because of problems in defining drought itself; the duration of a drought can vary, from months to years or even decades. However drought is defined, its effects can be particularly far-reaching and long term. Droughts are responsible for an estimated $6–8 billion in losses each year. In contrast, flood losses average about $2.41 billion per year, and losses from hurricanes tend to be $1.2 to $4.8 billion per year.

References

The American Meteorological Society and the University Corporation for Atmospheric Research. "Weather and Climate and the Nation's Well-Being." *Bulletin of the American Meteorological Society*, 73(12): 2035–2041, 1992.

Billion dollar U.S. weather disasters 1980–1999. National Climatic Data Center, 2000.

Environmental and Societal Impacts Group, National Center for Atmospheric Research. *Mesoscale Research Initiative: Societal Aspects Workshop Report*. Boulder, CO: Author, 1990.

Floodplain Management, <http://www.floodplain.org/p-learn.html>.

Hydrologic Information Center Flood Impacts Site. <http://www.nws.noaa.gov/oh/hic/>.

Impacts of Drought, from National Drought Mitigation Center, <http://enso.unl.edu/enigma/impacts.html>.

Knudson, T., and N. Vogel. "Superfloods Threaten West." *Sacramento Bee*, Nov. 23, 1997.

Knutson, C. *A Comparison of Droughts, Floods and Hurricanes in the U.S.* National Drought Mitigation Center, University of Nebraska, Lincoln. 2000. <http://enso.unl.edu/ndmc/>.

National Oceanic and Atmospheric Adminstration. *Worst Weather-Related Disasters of the Century.* 1999. <http://www.noaa.gov>.

National Weather Service Drought Information 2000. <http://www.nws.noaa.gov/om/drought.html>.

Pielke, R.A., Jr. "Who Decides? Forecasts and Responsibilities in the 1997 Red River Flood." *American Behavioral Science Review*, 7(2): 83–101, 1999.

"Storms Devastate Europe Timber." Reuters, Dec. 30, 1999.

"Toll from Storms in France Tops 80." Reuters, Dec. 30, 1999.

"Venezuelan Floods and Mudslides Caused by Environmental Factors?" *Gallon Environment Letter*, 4(1): 1–4, 2000.

"Venezuelan Foreign Minister Put Flood Toll at 10,000." CNN News, Dec. 20, 1999.

THE VALUE OF WEATHER INFORMATION

Public forecasts are clearly important and undeniably necessary. Despite, or perhaps because of, this necessity, debate has ensued over efforts to privatize forecasting services. In recent years, public forecasting services, including the Meteorology Service of New Zealand, the United Kingdom Meteorological Office, the Atmospheric Environment Program (AEP) in Canada, and the Bureau of Meteorology in Australia, have undergone unprecedented and often severe changes. Budget reduction efforts in their federal governments have been the main impetus behind these changes, which have forced forecasting services to privatize or self-finance. In the United States, forecast information and other weather data products from the National Weather Service are free. Private companies, such as AccuWeather or WeatherData Incorporated, can access these data and use them to provide the best possible information to their clients. Clients of this information—persons in transportation, utilities, insurance, newspaper, aviation, agriculture, emergency management, and other weather-sensitive industries—pay a fee to receive it.

Chuck Doswell and Harold Brooks at the National Severe Storms Laboratory address some of the issues relevant to determining the value of weather services, including the difficulties in obtaining a quantitative assessment of the value of weather forecasts. Forecast value cannot be simply related to reductions in the number of human casualties associated with severe or hazardous weather; calculations of the value of human life or the economic impacts of injuries and fatalities are not easy assessments to make. Establishing this value is especially essential in the face of budget cuts that may be imposed on the National Weather Service over the next five to ten years.

Attempts to establish the value of weather forecasts include analyses of how forecast information affects decision making at all levels. The decisions can be as simple as choosing to carry an umbrella or close a window, or as complicated as determining what and when to plant or whether to evacuate an area of coastline. A number of case studies have been examined dealing with these topics, including choices regarding frost prevention, irrigation, feed storage, and planting and harvesting. Weather and climate forecasts can also be important in decisions to generate electrical power. Utility companies can use this information to optimize production to meet the public's demands for heating or cooling. Detailed discussions and references to various case studies are provided in Chapter 4 of Katz and Murphy's *Economic Value of Weather and Climate Forecasts*.

Weather can inspire economic and even political and legal events in various ways. In April 1999, wine growers in southwestern France announced plans to sue Meteo France, the country's national weather service, for failing to predict hailstorms that devastated more than 49 square kilometers (19 square miles) of vineyards and 20 square kilometers (7.8 square miles) of fruit trees. Officials at Meteo France stated that forecasts had rated the risk of the storms at a level B on a scale of A to C; nevertheless, 800 of the farmers involved argued that adequate warning could have allowed them to protect their crops. Earlier in 1999, an unprecedented January snowstorm caused at least 300 kilometers (185 miles) of traffic jams in the Parisian Ile-de-France region and disrupted business in Paris's theaters.

Such events clearly demonstrate the importance of accurate and timely weather information. As we shall see, climate forecasts and information can also be relevant to decision making, especially when dealing with the issues of climate variability and climate change.

References
CNN Interactive, <http://www.cnn.com>.
Doswell, C., and H. Brooks. "Budget-Cutting and the Value of Weather Services." *Weather and Forecasting*, 13: 206–212, 1998.
Katz, R., and A. Murphy. *Economic Value of Weather and Climate Forecasts*. Cambridge: Cambridge University Press, 1997.

SOCIETAL IMPACTS OF CLIMATE VARIABILITY

Climate variability includes natural changes and oscillations like those related to El Niño and La Niña events. These events can significantly influence weather over large regions and can have dramatic societal and economic impacts. The 1997–1998 El Niño was the first such event to be accurately forecast. However, a verifiable prediction of El Niño or other climate variability does not necessarily correspond to accurate forecasts of the associated effects.

Michael Glantz at the Environmental and Societal Impacts Group at the National Center for Atmospheric Research has studied the effects and benefits of El Niño predictions on various industries and forecast users. One such industry is the Peruvian anchovy industry. The 1972–1973 El Niño devastated anchovy fishing in Peru to the extent that recovery was not possible until the 1990s. As a result of early forecasts of oncoming El Niño conditions, anchovy fishing in Peru was temporarily banned during April 1997, but perhaps due to industry pressure, the ban was lifted after just a few days. For the rest of 1997 and into 1998, catches plummeted, ultimately leaving many fisherman jobless.

In the case of El Niño, accurate predictions of the scale and intensity of the event, along with analysis of the potential impacts of these conditions, can help industries and governments prepare. On the flip side is reliance on forecasts despite an incomplete understanding of the limits of their certainty. This reliance can lead to overpreparation. The 1997–1998 El Niño climate forecasts called for drought conditions in Australia, causing many farmers to reduce the amount of agricultural planting. The dry conditions occurred, but the rain that did fall came at a crucial time to reduce agricultural losses. Those who had anticipated a more severe event lost out.

In any discussion of the effects of a climate phenomenon, the problem of attributability becomes an important issue. The El Niño of 1997–1998, for instance, seemed to take the blame for every weather-related event occurring anywhere in the world. Where the line should be drawn between what is and what is not directly attributable to an event or climate variability is not always clear. Current estimates hold the 1997–1998 El Niño responsible for some $8 billion in damages worldwide. The El Niño conditions have been linked to events in almost all regions of the planet, including

- flooding and landslides in California
- tornadoes in Florida
- snowfall in the southeastern United States
- flooding in Texas

- torrential rainfall and mudslides in Peru
- fires in Australia and Indonesia
- drought in New Zealand, coupled with flooding in many of the coastal regions
- severe storms bringing ten times the average annual rainfall to Chile
- five to ten times the normal amount of rainfall in Kenya and parts of Ethiopia
- flooding in Somalia
- rainfall deficits in Central America, the Caribbean, and northern South America
- an increase in tropical cyclone activity in the South Pacific
- monsoons in Pakistan and northwestern India
- severe summer heat in China and the Indian subcontinent
- widespread breeding of mosquitoes in South America
- drought in Papua New Guinea and Hawaii

El Niño and other climate anomalies can bring positive outcomes as well. Climatologist Stanley Changnon points out that atmospheric scientists and government agencies need "to focus on both good and bad impacts of predicted weather conditions." One assessment of the positive and negative impacts of the 1997–1998 El Niño concluded that the "prediction of 'bad impacts' raised major doubts about the scientists' predictions of negative outcomes apt to result from global warming" ("El Niño," 1998).

Table 3.1 summarizes the economic losses and gains associated with the 1997–1998 El Niño. The climate conditions associated with El Niño are credited with saving the United States almost $5 billion in heating costs, due to the mild winter weather in many parts of the country. Predictions of the warmer conditions enabled utilities to supply natural gas and heating oil at lower rates, which further reduced heating costs to consumers. In the Midwest and northeastern United States, temperatures nearly 7°C (12.6°F) above normal and little precipitation led to record-high consumer purchasing for January through March. The 1997–1998 El Niño led to the elimination of major Atlantic hurricanes in 1997, saving an estimated twenty lives (the average number of hurricane deaths per year since 1986). Furthermore, there were far fewer lives lost to extreme cold, winter snow

Table 3.1
Economic Losses and Gains Associated with the 1997–1998 El Niño

Economic Losses	Economic Gains
$2.8 billion in property losses	$6.7 billion in reduced heating costs
$400 million in federal government relief	$5.6 billion in increased sales of merchandise, homes, and other goods
$125 million in state assistance costs	$350–400 million in reduced costs of ice and snow removal
$650–700 million in agricultural losses	$6.9 billion in reduced losses due to an absence of snowmelt floods and reduction in Atlantic hurricanes
$60–$80 in lost sales of snow-removal equipment	$450–$500 million in income from increased construction and related employment
$180–$200 million in losses to the tourist industry	$160–$175 million in reduced operating costs to airline and trucking industries
Net Losses: $4.2–$4.5 billion 189 human lives lost	Net Gains: $19.6–$19.9 billion 850 human lives saved

Source: Compiled by Changnon (1999). Used by permission of the American Meteorological Society.

and ice storms, and weather-related vehicular accidents. The El Niño conditions also helped 1997 secure a position as the warmest year on record (a record that has since been broken; see Chapter 1).

Climate anomalies like El Niño and La Niña can affect precipitation patterns, and therefore weather in general, around the world. The effects of these weather changes on societies, and the ways in which public and private decision makers can best prepare for them, are subjects of much research. For instance, scientists have long known that climate has a natural variability: temperature, precipitation, and other variables can vary from year to year and exhibit large fluctuations over longer timescales. The frequency of extreme weather events, such as severe storms, can also vary from year to year and over longer time periods. Recently, there has been significant debate concerning whether the number of extreme weather events is increasing.

In a recent article, Kenneth Kunkel and coauthors note that "extreme weather events have led to ever larger economic losses in the

United States. Multibillion dollar losses now occur with increasing frequency." In analyzing whether these losses correspond to an increase in severe weather, the investigators examined trends in weather and climate extremes using records that date back as far as 1940. The analysis is tricky but seems to conclude that most of the data on extreme weather events do not show a correlative increase with time. The finding suggests that the increase in losses is primarily due to an increase in vulnerability arising from societal changes—for example, a growing population living in higher-risk coastal areas or large cities and lifestyle and demographic changes that subject lives or property to greater exposure.

Kunkel and his coauthors point out that more research is needed. Even something as seemingly simple as correlating increases in precipitation with changes in streamflow is fraught with debate. Much of the bibliographic material presented here contains many sources for information on climate variability and on fluctuations in weather and climate extremes. Other references listed address the political, economic, and other societal aspects of climate variability and the resulting effects.

References

Changnon, S.A. "Impacts of 1997–98 El Niño–Generated Weather in the United States." *Bulletin of the American Meteorological Society*, 80(9): 1819–1827, 1999.

"El Niño Prediction Uncertain—Lots More Known, But Not the 'Why'?" *Cincinnati Enquirer*, Mar. 26, 1998, p. A.15.

Glantz, M.H. "El Niño Forecasts: Hype or Hope?" *Network Newsletter*, 13, 1998.

Global Effects of El Niño. Environment Canada, <http://www.ec.gc.ca/envhome.html>.

Kunkel, K.E., R.A. Pielke, Jr., and S.A. Changnon. "Temporal Fluctuations in Weather and Climate Extremes That Cause Economic and Human Health Impacts: A Review." *Bulletin of the American Meteorological Society*, 80(6): 1077–1098, 1999.

National Aeronautics and Space Administration, <http://www.nasa.gov/>.

National Highway Traffic Safety Administration. *Data on Vehicular Deaths*. Washington, DC: U.S. Department of Transportation, 1999.

National Oceanic and Atmospheric Administration, <http://www.noaa.gov/>.

National Weather Service. *Storm Data. December 1997, January 1998, February 1998, March 1998*. Washington, D.C.: Government Printing Office, 1997–1998.

Pearce, J., and J. Smith. "Farewell to Winter That Wasn't: Warmth Hurt Businesses That Depend on Snow, But It Helped Consumers and Environment." *Detroit News*, Mar. 20, 1998, p. E.1.

Pfaff, A., K. Broad, and M. Glantz. "Who Benefits from Climate Forecasts?" *Nature*, 397: 645–646, 1999.

CLIMATE CHANGE IMPACTS

Related to weather and climate changes associated with climate variability are longer-term changes in our global environment. Over the years, scientists have observed and documented variations in temperature, rainfall, and other climate variables. But there is now concern, at least from a significant portion of the community, that the current trends point toward a warming earth and that the changes themselves may be speeding up or becoming more pronounced. Global warming and the controversy surrounding it are discussed in detail in Chapter 1. In this chapter, we focus on some of the possible effects of such changes on our societies, including the effects to human health and livelihood.

World Food Supply

The world's food supply is derived from crops or from animals that consume crops. Agriculture therefore plays an important role in the continued sustainability of our population. Crop production is very sensitive to extremes in temperatures, precipitation, and other climatic variables. This sensitivity was studied recently in an international effort led by Cynthia Rosenzweig, a research agronomist at the Goddard Institute for Space Studies in New York City. Analysis of various biophysical and ecosystem models found large differences in agricultural vulnerability to climate change, depending mainly on the latitudinal effects of global warming and the response of particular plants to increased carbon dioxide in the atmosphere. Knowledge of many interrelated factors is essential to reliable estimation of the impacts of climate change on food supply, yet the process is not without some debate or uncertainty. Sources of uncertainty and question are the degree and geographical distribution of temperature increases, changes in precipitation and evaporation rates due to a warmer climate, and the physiological response of crops to a carbon dioxide-enriched atmosphere.

Many are quick to point to documented evidence that a warmer climate may offer some advantages to crop production. One of these advantages is increased carbon dioxide assimilation. Plants are categorized into physiological classes depending on their pathways for carbon dioxide uptake. In experimental studies in greenhouses and other controlled environments, C_3 plants,—wheat, rice, and soybeans—responded favorably to increased carbon dioxide levels. C_4

plants—corn, sorghum, millet, and sugarcane—are highly efficient in their utilization of carbon dioxide at current concentrations but tend to be less responsive in an enriched atmosphere. In either case, a carbon dioxide–enriched environment may improve water use efficiency, defined as the ratio between a crop's biomass and the amount of water consumed. Another potential advantage for crop production in a warmer climate is an extended growing season at mid- and high latitudes. A lengthening of the growing season has already been observed in some regions and is directly attributed to changes in air temperature. Analysis of over thirty years of data by Anne Menzel and Peter Fabian at the University of Munich shows that the average growing season in Europe has lengthened by 10.8 days since the early 1960s. Their report indicates that the onset of spring events such as leaf unfolding would advance up to six days per 1°C (1.8°F) increase in the winter air temperature. Warmer temperatures allow earlier planting and faster maturation times, thus opening the possibility of two or more cropping cycles each season. Such increases in temperature could also extend the crop-producing areas northward into Canada and Russia. A shift in precipitation patterns could mean increased water availability, though such effects would certainly be only regional.

Despite these possible advantages to agricultural productivity, a number of drawbacks must be anticipated as well. Changing precipitation patterns may leave many areas more susceptible to drought, especially as warmer temperatures subject plants to increased heat stress and soil evaporation. In addition to damage due to water shortages, increased evaporation may lead to accelerated erosion. This effect, combined with the prospect of flooding due to a rise in sea levels, would raise the salt content of the soil, thus reducing its fertility. The faster growth mentioned as a benefit might also have a negative side effect realized in lower yields, and a warmer climate is likely to be more favorable to many pests and plant diseases.

Calculated changes in wheat yield for an increased carbon dioxide scenario vary by region. Overall, the increased heat and water stresses at low latitudes are predicted to cause a decrease in agricultural production, while the lengthened growing season and warmer summers would result in an increase in food production at mid- and high latitudes. Such shifts in agricultural production could mean large changes in world food policies. Results by Rosenzweig and collaborator Martin Parry of Oxford University show an 11 to 20 percent reduction in worldwide agricultural production due to global warm-

Figure 3.3. An Alabama farmer looks at his drought-stricken corn crop. Variations in climate can seriously affect crops through drought, disease, and changes in the location of prime agricultural areas. These changes could have implications for food supply. © Reuters Newmedia Inc./CORBIS.

ing. The results also indicate that with adaptations by farmers and government investments in infrastructure, this change in production could be limited to -5 to $+1$ percent.

Adaptations by farmers include introducing later-maturing crop varieties, switching crop sequences, adjusting the timing of planting and other operations, using tillage methods that conserve moisture, and improving irrigation efficiency. Economic and other infrastructure-type modifications may include shifts in regional production centers; reallocations of land, capital, and labor; genetic engineering of heat- and drought-resistant crop varieties; or increasing the harvest index, which is the part of the total plant matter that is marketable. Such adaptations are limited by current agricultural and socioeconomic realities, however. As Rosenzweig and Hillel point out, changes in crop variety or time of harvest may expose farmers to market problems or credit crises associated with higher capital and operating costs. Low-latitude countries are in many cases also lower-income countries; resources for adaptation may be unavailable. Water plays a crucial role as well; even with adaptations, there are limits on the availability and viability of irrigation. Also, there is currently no evidence indicating that crop modification can ensure equal levels of food production or nutritional quality. Other detriments associated with less understood effects of climate change may also prove significant. An example is the case of droughts or floods disrupting the transport of grain on rivers.

Natural climate variability plays a significant role in predictions of the future environment. British climate researcher Mike Hulme and coauthors completed a study of the impacts of this unpredictability of the climate system. Hulme's group examined multicentury simulations of a nonperturbed climate in conjunction with an ensemble of increased-greenhouse-gas-scenario simulations to determine the effects on river runoff and wheat yield. Their results indicated that human-induced climate change impacts on mean river runoff by the year 2050 would be significantly greater than natural climate variability in northern and southern Europe. In central and western Europe, however, the effects were predicted to be less than those associated with natural climate variability. Wheat productivity showed very high sensitivity to natural climate variability, with production varying about 8 percent of the mean 1961–1990 yield in Italy and about 14.6 percent in Romania. The models indicated impacts attributable to climate change alone in the countries of Finland, Germany, and the Netherlands, and in these cases, the effect was an increased wheat yield. Elsewhere, changes in wheat yield due to global warming were virtually indistinguishable from those due to climate variability. Such results allude to the difficulties associated with detecting regional-scale climate change impacts. They also suggest that adaptations to natural climate variability may provide a medium-scale defense against the effects of global warming in some locations, though in others such adaptations may not be nearly enough.

Without doubt, studies indicate that the Intergovernmental Panel on Climate Change's predicted temperature increase of 1.5 to 4.5°C (2.7 to 8.1°F) will directly affect the food-producing capability of most nations. Preliminary findings indicate that the impacts will be greatest for lower-latitude countries where rapidly growing populations can least afford a decrease in production or the financial cost of adaptation. The many uncertainties relating climate change and agriculture point to the necessity of more research and collaboration by economists and physical and social scientists. Only then will policymakers and the public have the ability to make the most effective decisions regarding agricultural sustainability.

References

Burroughs, W.J. *Does the Weather Really Matter? The Social Implications of Climate Change.* Cambridge: Cambridge University Press, 1997.

Hulme, M., E.M. Barrow, N.W. Arnell, P.A. Harrison, T.C. Johns, and T.E. Downing. "Relative Impacts of Human-Induced Climate Change and Natural Climate Variability." *Nature,* 397: 688–691, 1999.

Menzel, A., and P. Fabian. "Growing Season Extended in Europe." *Nature,* 397: 659, 1999.

Rosenzweig, C., and D. Hillel. "Potential Impacts of Climate Change on Agriculture and Food Supply." *Consequences,* 1(2): 23–32, 1995. <http://www.gcrio.org/CONSEQUENCES/introcon.html>.

Rosenzweig, C., and D. Hillel. "Agriculture in a Greenhouse World: Potential Consequences of Climate Change." *National Geographic Research and Exploration,* 9: 208–221, 1993.

Rosenzweig, C., and M.L. Parry. "Potential Impact of Climate Change on World Food Supply." *Nature* 367: 133–138, 1994.

Ruttan, V.W., ed. *Agriculture, Environment, Climate and Health: Sustainable Development in the 21st Century.* Minneapolis: University of Minnesota Press, 1994.

Smith, J., and D. Tirpak, eds. *The Potential Effects of Global Climate on the United States. Report to Congress.* Washington, DC: U.S. Environmental Protection Agency, 1988.

Tegart, W.J. McG., G.W. Sheldon, and D.C. Griffiths, eds. *Climate Change: The IPCC Impacts Assessment.* Canberra: Australian Government Publishing Service, 1990.

U.S. Congress, Office of Technology Assessment. *Preparing for an Uncertain Climate.* Washington, DC: U.S. Government Printing Office, 1993.

Damage to Human Health

Research has shown that a warming of the earth's climate could lead to an increase in the spread of tropical diseases such as malaria. As reported by Paul Epstein of Harvard University's Medical School, changes in regional climate and land development have influenced a quadrupling of the number of malaria cases in just five years. About 2 billion people in the tropics and subtropics are at risk, and each year about 500 million new cases are reported worldwide. Mortality is significant only among the very young, causing 1.5 to 3 million deaths per year, but it may increase if malaria spreads to new areas. Malaria is an insect-borne disease, transmitted by the anopheles mosquito. This mosquito can survive only where the winter temperature remains above 16°C (61°F) and is found in Africa, Asia, and North and South America. Its relative, *Aedes aegypti,* which carries dengue fever, cannot survive in regions where winter temperatures are below 10°C (50°F). In recent years, both of these insects have been reported at higher elevations, particularly in the African highlands and in Latin America and Papua New Guinea. As a warming climate causes isotherms to shift poleward and to high elevations, climate models indicate up to a 50 percent increase (25 million square kilometers or 15.5 million square miles) in the extent of the area susceptible to endemic or seasonal malaria. Such an increase could encompass many parts of Europe, Russia, North America, and Australia, resulting in 50 to 80 million new malaria cases each year, with a mortality rate as high as 20 percent.

Floods and droughts may also play a direct role in the spread of epidemics. As reported by Thomas Karl and coinvestigators at the National Climatic Data Center, the years since the 1970s have seen an increase in the number and duration of extreme weather events in the continental United States. Such extremes mean longer droughts, greater likelihood of fires, more frequent occurrences of sudden, intense precipitation, or more severe cold weather periods, all of which may directly affect health. Drought conditions can reduce the number of prey animals, thereby favoring population explosions of insects and rodents. This situation occurred in the southwestern United States in the early 1990s when a prolonged drought decreased the populations of coyotes, owls, snake, and other predators. The result, when combined with an increase in food supply following intense rains, was a tenfold increase in the rodent population by June 1993. By the end of summer, numerous cases of a "new" disease, hantavirus pulmonary syndrome, spread by rodent saliva and droppings, had been reported. Such hantaviruses have also emerged in several European nations, Canada, and South America. Drought is also named a factor in the record meningitis outbreak in West Africa in 1996, in which over 100,000 persons were affected and more than 10,000 died.

A warming of our climate could result in more intense heat waves, also contributing to an increase in mortality. A large number of deaths in Chicago and other major cities worldwide during the summer of 1995 were directly attributed to heat waves, with the key factor being lack of relief at night. Meteorological data indicate that minimum daily temperatures (corresponding to the nighttime low) have risen at a rate of 1.86°C (3.35°F) per 100 years for the latter half of this century. In contrast, maximum daily temperatures have risen at a rate of 0.88°C (1.58°F) per century. A 2–4°C (3.6–7.2°F) rise in summer temperatures would mean more days with temperatures above 35°C (95°F) and could increase the number of heat-related deaths each year in cities like New York or Chicago by several hundred. In developing countries where air-conditioning is not widely available, the mortality rate would increase even more dramatically.

The increase in ultraviolet (UV) radiation reaching the earth's surface as a result of ozone depletion in the stratosphere is also a source of concern. The ozone layer acts to block UV radiation from the sun. UV radiation has wavelengths just shorter than the wavelengths of visible light. In small quantities, UV can initiate the production of vitamin D3 to build and maintain our bones. In larger quantities, the effects can be quite damaging. The human skin has developed various defense mechanisms to counteract the effects of increased UV exposure, including thickening the epidermis or outer layer and developing pigmentation to shade more vulnerable and deeper cells. When these natural protections are overcome, molecular damage to cells, as well as an inability to repair this damage, may result.

Damage from UV is linked to an increased occurrence of nonmelanoma skin cancers, which are among the most frequently diagnosed and rapidly rising forms of cancers in white populations. About 600,000 cases are diagnosed each year in the United States alone, and increased UV exposure due to decreased ozone is anticipated to result in an extra 100,000 cases per year by the middle of the twenty-first century. Melanoma, a cancer of the pigment cells, is rarer, affecting about 17,000 men and 12,000 women in the United States each year. Mortality is as high as 20 percent for melanoma cases, compared to about a 1 percent mortality for nonmelanoma cancers. Studies since the mid-1980s have suggested the possibility of a link between melanoma and UV overexposure, although quantification of the exact impact of ozone depletion on melanoma has been difficult to determine.

Exposure to UV radiation has also been linked to eye damage, causing vision reduction and, in some cases, blindness. Effects on the eye include pterygium (an outgrowth on the outermost layer of the eyeball), keratopathy (a degeneration of the fibrous layer covering the lens), and cortical cataracts. Tentative estimates indicate about a 0.5 percent increase in cataract incidence for each 1 percent decrease in average ozone concentration. Given recent measurements of a 6–7 percent ozone decrease in the summertime, this would result in a 3 percent increase in cataract incidence over the next several decades.

References

Buchmann, S.L., and G.P. Nabhan. *The Forgotten Pollinators.* Washington, DC: Island Press, 1997.

Burroughs, W.J. *Does the Weather Really Matter? The Social Implications of Climate Change.* Cambridge: Cambridge University Press, 1997.

Epstein, P.R. "Climate, Ecology, and Human Health." *Consequences,* 3(2): 3–19, 1997. <http://www.gcrio.org/CONSEQUENCES/vol3no2/>.

Garrett, L., ed. The Coming Plague: *Newly Emerging Diseases in a World Out of Balance.* New York: Farrar, Strauss, and Giroux, 1994.

Godfrey, C. "Climate Change and Human Health." <http://www.ncdc.noaa.gov/pw/cg/humh_doc.html>.

Grifo, F., and J. Rosenthal. *Biodiversity and Human Health.* Washington, DC: Island Press, 1997.

de Gruijl, F.R. "Impacts of a Projected Depletion of the Ozone Layer." *Consequences,* 1(2): 13–21, 1995. <http://www.gcrio.org/CONSEQUENCES/vol1no2/>.

Karl, T.R., R.W. Knight, D.R. Easterling, and R.G. Quayle. "Trends in U.S. Climate during the Twentieth Century." *Consequences,* 1(1): 3–12, 1995. <http://www.gcrio.org/CONSEQUENCES/spring95/>.

McMichael, A.J., A. Haines, R. Slooff, and S. Kovats, eds. *Climate Change and Human Health.* Geneva, Switzerland: WHO/WMO/UNEP, 1996.

Peters, R.L., and T. Lovejoy. *Global Warming and Biological Diversity.* New Haven, CT: Yale University Press, 1992.

President's Office of Science and Technology Policy and IOM. *Human Health and Climate Change.* Washington, DC: Government Printing Office, 1996.

World Health. *Report 1996: Fighting Disease, Fostering Development.* New York: World Health Organization, United Nations, 1997.

Wyman, R.L., ed. *Global Climate Change and Life on Earth.* New York: Routledge, Chapman and Hall, 1991.

Economic and Political Concerns

Global warming is a global issue. It will affect and must be addressed by all countries of the world. The myriad of factors involved in international negotiations make understanding this issue a complex task. Potential areas of disagreement are due in part to the fact that developing countries have less infrastructure in place to mitigate the effects of climate change and may therefore be more severely affected. In implementing greenhouse gas reduction strategies, large countries relying significantly on coal and fossil fuels would be forced to bear the bulk of the greater part of the economic burden. The issue is further complicated because China and India are major contributors of sulfate aerosol, which some investigators theorize will partially offset the warming effect.

The potential costs of global warming–induced food shortages, property damage, or health risks must be weighed against those of reducing emissions of greenhouse gases to the atmosphere. Economist William Cline completed a survey of a number of economic models of climate change and concluded that all of the models showed the costs of reducing carbon emissions to be relatively modest and of the same order. According to this study, reduction of emissions of 75 percent by 2100 would mean a decrease of only 2 to 4 percent in the worldwide gross domestic product (GDP), a measure of the market value of an economy's produced goods and services. This amount is the same order of magnitude as the estimates of damages caused by global warming. Still, considering that the worldwide economy is worth an estimated $20–30 trillion, a 2 to 4 percent reduction in the GDP is not small change. It is not surprising then that potential influences and impacts of climate change should be topics of much debate and contention.

In December 1997, delegates from 159 countries met in Kyoto, Japan, in an attempt to reach an international agreement on an approach to climate change. Known officially as the Third Meeting of the Conference of the Parties to the Framework Convention on Climate Change (COP-3), the congress included representatives from countries that had signed the original Climate Change Convention at the 1992 Earth Summit. The goal of the meeting was to formulate a protocol obliging the thirty-nine industrialized countries to reduce their greenhouse emissions according to fixed targets and timetables. This net reduction of carbon dioxide, methane, dinitrogenoxide, hydrofluorocarbon, perfluorocarbon, and sulfur hexafluoride emissions would represent a trend reversal of over 100 years of reliance on fossil fuels.

From the beginning, negotiations were plagued by differences between developing and industrialized countries. In the early morning hours well after the conference's scheduled end, an agreement was finally reached. Reduction goals were established for each of the major industrialized countries; 132 developing countries were exempted from the treaty. To some, the Kyoto outcome represented a historic success; to others, including scientists as well as politicians, it was a source of much disagreement. Many believed the reductions were not enough; others suggested that the economic repercussions may not be worth any possible gain, and no one seemed happy with the developing countries' exemptions. Atmospheric scientist Vinod Saxena states, "With but fragmentary knowledge at hand, there is no evidence anywhere that the enforcement of the Kyoto protocol will cause any reduction in global temperatures. On the contrary, it certainly will cause major economic dislocations, particularly in the regions where coal-fired power plants, coal mining and fossil fuel–related industries are located." In summarizing an analysis of fossil fuel emissions through the end of the twenty-first century, Stephen Schneider, a long-time researcher of climate change and participant at Kyoto, says, "Clearly, much more needs to be done." Both voice the thoughts of many individuals, and both have much evidence to support their views. In the scope of such controversy, attaining global agreement seems essentially impossible.

The Fourth Meeting of the Parties took place in Buenos Aires in November 1998 and brought the scale and complexity of climate change and politics to a head. After two weeks of deliberations, the United States and China remained locked in a debate over climate policy that was expressed to "at times, rival the bitterness of the Cold War." In the end, the mandates of the Kyoto Protocol were upheld, although the treaty still required ratification, which—if not completed in three to four years—would make any action too late to meet the established goals. In general, the response to Buenos Aires seems as mixed and perhaps as joyless as that to Kyoto. According to one participant, Worldwatch Institute's Christopher Flavin, "Buenos Aires has not yet produced the 'good air' that its name implies—and it will take a lot more political leadership in the year ahead to ensure that it was even a small step in that direction" (Buenos Aires Climate Negotiations). In spring 2001, the Bush administration announced the decision that the United States would not ratify the Kyoto Protocol.

The Kyoto legislation aimed to counter global warming by means of mitigation: that is, preventing future climate impacts on society through the limitation of greenhouse gas emissions. The effects of these limitations on the total amount of carbon dioxide in the atmosphere are shown in Figure 3.4. Problems with compliance and other limits on the legislation make it unlikely that mitigation efforts will be sufficient for attaining their goals. As pointed out by Roger Pielke, Jr., a political scientist at the National Center for Atmospheric Research, there is a very real possibility that mitigation efforts will not succeed, and even if they do, maintaining current standards of living will require additional measures.

The additional measures involve adaptation: adjustments in individual, group, and institutional behavior in order to reduce society's vulnerability to climate variations or changes. According to Pielke, adaptation is not a single response; it should be a portfolio of re-

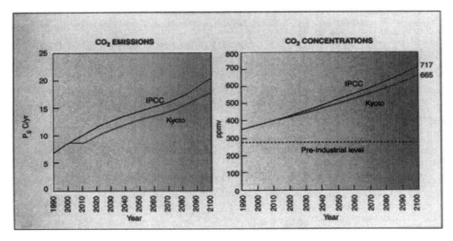

Figure 3.4. Effect of the Kyoto Protocol on (left) the projected total worldwide emission of carbon dioxide (CO_2) and (right) the resulting concentration of gas in the atmosphere (in parts per million per volume, ppmv), through the next 100 years. Emissions are in units of petagrams (10^{15} grams) of carbon, or equivalently, billions (10^9) of metric (2200 lb) tons per year. In each figure, the reference line (labeled IPCC) is the projected scenario, called IS92a, developed by the Intergovernmental Panel on Climate Change. Shown for reference in the second figure is the preindustrial CO_2 concentration of roughly 275 ppmv. From J.A. Edmonds, "Beyond Kyoto: Toward a Technology Greenhouse Strategy," *Consequences*, 5(1), 1999.

sponses, and the responsibility for implementing these responses should be shared. As with mitigation, however, the process of implementing adaptation is plagued with problems. There are problems with costs and determining who should bear the burden. Proponents of adaptation strategies point out that adaptation is beneficial in any context and should be viewed as more than a complement to mitigation efforts, as suggested by the Intergovernmental Panel on Climate Change. The debate continues, and current events suggest the controversy is not likely to be resolved any time soon.

References
Buenos Aires Climate Conference: Fourth Conference of the Parties, <http://www.greenpeace.org/luxembrg/klimakonf-eng.html>.
Buenos Aires Climate Negotiations. Worldwatch Briefing, <http://www.worldwatch.org/alerts/981113.html>.
Burroughs, W.J. *Does the Weather Really Matter? The Social Implications of Climate Change*. Cambridge: Cambridge University Press, 1997.

Cline, William R. *The Economics of Global Warming.* Insititute for International Economics, Washington, DC, 1992.

From Kyoto to Buenos Aires: A Summary of the Kyoto Meeting and Expectations for the Upcoming Meeting in Buenos Aires. <http://www.oneworld.org/earthaction/en/Recent/98-09-C1Ch/kyoto.html>.

"Harsh Climate: Congress Cool to Kyoto Protocol." *Global Change*, 4(1): 2–3, 1998.

"The Kyoto Protocol—A Milestone on the Approach to Sustainability." *WMO Bulletin*, 47(2): 161–163, 1998.

Pielke, Roger A., Jr. "Rethinking the Role of Adaptation in Climate Policy." *Global Environmental Change*, 8(2): 159–170, 1998.

Saxena, V. Hysteria Overtakes Good Science. High Point, NC: High Point Enterprise, 1998.

Saxena, V. Interviewed in High Point Enterprise High Point N.Y. Sept. 22, 1998.

Schreider, S. "The Global Warming Debate." Lecture presented at the University of Colorado at Boulder, Mar. 1999.

WEATHER MODIFICATION

The previous sections addressed how weather and climate affect society and how human beings can modify and will perhaps need to adapt to climate. We must not leave out the ways in which society can affect weather. In 1946, Vincent Schaefer and Irving Langmuir of the General Electric Research Laboratory found that dropping dry ice into a cloud could quickly lead to the production of snowflakes and the dissipation of the cloud in the seeded region. Their work marked the beginning of modern weather modification.

Weather modification can be thought of as the use of human intervention to relate the weather to human purposes. The term applies to the intentional treatment of individual clouds or storm systems to influence or improve the atmospheric processes that constitute weather conditions. The goal of such treatments is to achieve a beneficial effect without harming people or the environment. Weather modification most commonly applies to attempts to augment rainfall, dissipate fog, or reduce the size or amount of hail from severe storms. These applications can be of tremendous value to the agriculture and aviation industries.

The most common weather modification process, known as cloud seeding, uses chemical materials to alter the physical processes leading to the formation and growth of cloud water droplets and ice crystals. This can enhance precipitation and may be able to suppress hail formation. Extensive scientific experiments and the application of scientific concepts have led to the development of many new weather modification techniques. This research has focused on obtaining a better understanding of snow, lightning, and devastating winds, and in some cases, the techniques have been reported to provide very predictable results.

Weather modification activities tend to fall into one of three categories. The first involves the injection of energy by brute force, such as the use of powerful heat sources or intense mechanical mixing of air by fans or helicopters. These methods are used mainly in situations involving fog dispersal. The second category involves the alteration of land and water surfaces in an attempt to modify natural interactions with the lower atmosphere. Techniques in this category include blanketing the land surface with a dark substance to increase absorption for stronger surface heating, which leads to greater convection, forming more clouds and eventually rain. The third category of weather modification techniques includes attempts to intensify or redirect the atmosphere's natural energies. Cloud seeding, which alters the physical processes of the cloud, is an example.

Historically, weather modification has had a somewhat controversial reputation in the meteorological community. Many feel that interference with natural processes may lead to unpredictable and potentially detrimental effects. Some weather modification ideas, such as hurricane routing, pose interesting political questions, although the ability to route hurricanes does not seem likely to occur at any time in the foreseeable future. In many aspects, weather modification has become mainly the realm of private industry. Seeing the economic potential of being able to influence changes in precipitation, the commercial sector was quick to take hold of and develop this new technology. Eventually, budget considerations led to the demise of many experimental cloud seeding programs. These programs were simply unable to compete with the immediate reward promised by operational uses. The development of weather modification techniques made significant contributions to cloud physics research, however, and over the past few decades, much of this research has been linked, often indirectly, to obtaining an improved understanding of the risks and reliability of weather modification techniques.

Outside of the industries most directly affected by weather—most notably aviation and agriculture—little may be known concerning weather modification. For many, the weather may simply be what it is. Rain is rain, and fog is fog, and if the weather interferes with our picnic plans, we can try again another day. But for some things bigger than picnics, another day might not be good enough or even possible.

Airports often rely on weather modification techniques to disperse fog on runways, although not all fogs can be successfully dissipated. Fruit growers use heating lamps and other techniques to protect their crops on chilly evenings. By heating, stirring, or otherwise altering the air near the surface, they are in essence modifying the natural state of the weather.

A good example of a major weather modification client is the U.S. Air Force, which has used these techniques for a number of years for dispersing fog on its bases. The method involves spreading dry ice pellets or silver iodide particles into layers of supercooled fog or low cloud. The particles trigger the formation of ice crystals, which then grow at the expense of water droplets. The result is a snow shower that opens a large hole in fog or cloud and significantly improves visibility. The ice crystal technique does not work on warm fogs, however. In these situations, a mixing mechanism (often a helicopter) is used to stir in drier, warmer air, or attempts are made to heat and evaporate the fog. Commercial airports may use weather modification techniques as well. Orly Airport in Paris has a sophisticated system of eight engines located in underground chambers along the upwind edge of the runway to be used for this purpose. The system, called Turboclair, was installed in 1970 and is capable of improving visibility for distance of about 900 meters (almost 1000 yards).

Frost prevention, of extreme importance to fruit growers, is usually accomplished by reducing the amount of heat lost during the night or adding heat to the lowermost layer of air. To minimize heat loss, growers cover plants with paper, cloth, or other material having a low thermal conductivity. Smudge pots, producing smoke and soot particles, can be used to suspend particles in the air and help hold heat near the surface. Other methods involve sprinkling the plants with water, which will itself add warmth and release fusion energy (heat) when it freezes, or using a wind machine to mix air aloft with colder surface air. The latter method is successful only when the temperature 15 meters (50 feet) above the ground is at least 5°C (9°F) warmer than the surface. Surface heating of a fruit grove requires about thirty to forty heaters per acre, corresponding to substantial fuel costs. Still, modifying the weather in this way is often a much better solution than losing an entire crop.

Another area of potential weather modification applications is hail suppression. Hail is a major cause of crop losses and property damage. Experiments to reduce the size or amount of hail from a thunderstorm have been tried for a number of years, but reported successes have been minimal. In the Soviet Union in the 1960s, scientists used rockets and artillery shells to carry freezing ice particles into clouds. The theory was that the additional ice particles would compete for the cloud's liquid water supply, and any hailstones would not be able to grow large enough to cause destruction. There is little documentation concerning the results of these experiments. In the 1970s, U.S. researchers undertook a National Hail Research Experiment in northeastern Colorado. Analysis of three years of data showed no statistically significant difference between seeded and nonseeded incidences of hail, and the planned five-year experiment was terminated early.

References
Atmospherics Incorporated Weather Modification, <http://www.atmos-inc.com/weamod.html>.
Changnon, S.A. "Where Does Weather Modification Fit Within the Atmospheric Sciences?" *Journal of Weather Modification*, 24(1): 118–121, 1992.

SUMMARY

Humans are affected by weather. Our infrastructure is constructed in such a way that it is vulnerable to storms, floods, and other weather events. Climate, and possibly climate changes, can also have an impact on our society. Changes in weather and climate can influence health, the food supply, economic systems, and general well-being. Humans can also influence the weather, as demonstrated by weather modification techniques for seeding clouds, dispersing fog, and preventing frost. Humans may be influencing climate by putting increased amounts of substances, some natural and some of our own invention, into the atmosphere. In general, the ties between weather and climate and society are complicated. The aspects of meteorology that directly and indirectly influence society are myriad. The questions are many, and absolute answers are few. Chapter 4 will outline some of the unsolved problems in atmospheric research, but for many meteorologists and social scientists, the biggest pieces of the puzzle are the ones relating the intricacies of our atmosphere and society and addressing the impacts of each on the other.

Chapter Four

Unsolved Problems, Unanswered Questions

The atmospheric sciences are a young science; most of the current state of knowledge was developed in the twentieth century and in the later part of the century at that. Still, the amount of knowledge accumulated on various atmospheric topics has grown enormously. Chapter 1 provides an overview of recent and ongoing research in many areas of meteorology. This chapter provides a more detailed look at topics that are less well understood. Meteorologists and atmospheric scientists have identified these topics as areas in which further research should be a necessary priority.

In general, the unanswered questions in the meteorological sciences have become more focused in recent years. As knowledge of atmospheric processes grows, for example, the questions move toward attaining a better understanding of particular mechanisms that drive climate, while in the past a significant research challenge may have been simply identifying those mechanisms.

Understanding climate change and variability continues to be an important topic in twenty-first century-atmospheric science. The many outstanding issues in climate research are discussed in greater detail below and are among the most complex questions currently under examination. As described in Chapter 1, there is still debate within the scientific community about global warming and natural climate variability. Future work is likely to try to elucidate the role of both natural variability and human impacts in governing climate. This work is truly interdisciplinary, requiring a thorough understanding of land, ocean, and atmospheric processes and of the interactions among all three.

To answer some of the questions of meteorology, improved methods for observing the global atmosphere, lands, and oceans are considered imperative. These methods include new observation capabilities that can resolve meteorological variables on space and time-scales that are relevant to meaningful local and global forecasts. The observations would help resolve land-ocean-atmosphere interactions, as well as interactions among the different atmospheric scales. Integration of these global observations can help produce advances in weather forecasting and air quality.

Advances in meteorological understanding, combined with increased observation capabilities and developments in computer technology, are likely to result in improved numerical weather prediction techniques. These techniques are applicable not only to weather forecasting but also to issues in atmospheric chemistry, air quality, climate, and space weather research. Improved capabilities for prediction can be of great value and service to society and will improve links between weather and climate issues and human concerns. For instance, improved climate predictions can be used to help guide water resource management and will help better define the implications of rapidly increasing greenhouse gases.

Weather forecasts of the future are likely to improve along two different directions. First, the spatial scales for which information is available are likely to become smaller. The largest benefits of such local-scale predictions will be in the forecasting and warning of severe storms. Improvements at this level can directly translate to lives saved and property protected. Second, advances in the weather prediction models should translate to improved forecasts further in advance. Although chaos theory dictates that the upper limit on accurate, detailed weather prediction may be about two weeks, current forecast models are usually considered accurate only to about five to seven days. Future developments in forecasting should extend this range to seven to fourteen days and should improve the accuracy of predictions at longer timescales.

The following sections highlight many of the outstanding questions in the various subdisciplines of meteorology. Although the material is written for a general audience, some of the information here tends to be more specialized than elsewhere in the book. While significant effort has been made to ensure that the topics are accessible, the following sections may present information at a slightly more advanced level. Readers less familiar with the language of meteorology may benefit by reading Chapter 1 first.

FUTURE DIRECTIONS IN ATMOSPHERIC DYNAMICS RESEARCH

Atmospheric dynamics, or dynamic meteorology, is the study of the motions of the atmosphere. These motions range from gentle breezes to movements of large-scale systems and are closely associated with the weather and climate of a region and of the globe. Meteorologists have examined atmospheric dynamics on a variety of scales, from larger-scale synoptic (scales of hundreds of kilometers or greater) systems to smaller-scale storms (on scales of a few up to hundreds of kilometers). Improving what scientists know about atmospheric motions has helped create a solid scientific foundation for weather forecasting. Further improvements in understanding these motions can lead directly to more accurate and timely weather predictions and warnings. Future atmospheric dynamics research is likely to focus on the issues outlined below.

Development of better representations of small-scale physical processes. Smaller-scale features, including localized areas of heavy rain or winds, are often embedded in larger-scale weather systems. The inability to forecast the formation and movement of these smaller-scale features leads to difficulties in producing accurate precipitation forecasts. Specifically, the effects of radiation and latent heating on the formation of fronts and weather systems must be better understood. Since smaller-scale weather systems can instigate weather tens of kilometers away from their location, knowledge of the interactions of smaller-scale and larger-scale processes must be improved.

Improved measurements of atmospheric conditions associated with hurricanes, including pressure differences, winds, and precipitation amounts. Predictions of hurricane intensity are often inaccurate, yet these predictions are extremely valuable to protection of life and property. Research shows that physical processes in the hurricane boundary layer and the upper levels of the troposphere can significantly influence hurricane intensity. These processes include interactions between the air and the sea surface, especially at high wind speeds. Hurricanes and tropical storms that make landfall can cause major inland flooding, leading to significant property damage and a potential loss of life. The dynamics associated with landfalling tropical cyclones are not well understood; this topic has been recommended as an area of focus for future research.

Better understanding of intense, short-lived downward winds, called downdrafts or microbursts. These phenomena are associated with convective storms and remain an area of great research interest and practical concern, in part because of the threat they pose to aircraft. The 1994 crash of a U.S. Air jet on approach for landing in Charlotte, North Carolina, was attributed to a possible downdraft. Downdrafts are difficult to detect due to their localized nature and short lifetimes.

Improved water vapor measurements to aid in predicting when and how a storm will form. Atmospheric water vapor can vary significantly with space and time, and its distribution is generally not well measured. Measurements of soil moisture, which strongly influence atmospheric boundary layer processes over land, are also severely lacking. Routine measurements of soil properties, coupled with higher quality and quantities of atmospheric water vapor estimates, could provide necessary information to improve predictions of storm initiation.

Better understanding of the limits of seasonal forecasts. Seasonal predictions of weather associated with specific phenomena such as El Niño have resulted in some success, but many questions remain concerning the seasonal prediction process. For instance, what mix of statistical methods will provide the most accurate seasonal predictions, and how accurate might these predictions be? What is the influence of initial conditions, and what methods are most useful in obtaining global measurements of these conditions?

Improved understanding of the influences of geography and topography, including how mountain ranges or other terrain affect winds and precipitation. The earth's terrain can influence many types of atmospheric phenomena, including modifying fronts and thunderstorms, enhancing or decreasing rainfall, and blocking or intensifying strong winds. Many deserts, for example, are located just downwind of a mountain range, in what is known as a rain shadow. Variations in land surface properties can affect the thermodynamics and water vapor content of the overlying atmosphere and play an important role in regional climate. Accurate simulations of the effects of geography and terrain in weather forecast models require more precise measurements of land surface properties and the ability to capture these features at appropriate spatial scales. The models would also need to be able to resolve the physical processes associated with these smaller scales more accurately.

Improved understanding of large-scale teleconnection patterns. Teleconnections are linkages between weather patterns in widely separated parts of the globe and can help scientists explain why changes in one part of the world can affect weather patterns halfway around the globe. Teleconnection patterns play a large role in determining the effects of El Niño on weather and climate worldwide. Currently, the structures of these patterns, along with the roles of atmospheric instabilities and irregular whirls or rotational movements called eddies, are not well understood. Other questions concern the limits on predictability associated with these patterns and the causes of these limits, as well as the role of the stratosphere in teleconnection patterns.

Improved understanding of the structure and variability of storm tracks and eddies. Many atmospheric motions are treated as *linear* disturbances in models, meaning that they are represented by linear equations able to yield a single solution. The atmosphere itself is highly *nonlinear*, however, and its motions are most accurately represented by nonlinear equations that are often insolvable; the solutions can only be approximated. When the approximations break down, turbulence is the part that is left. Scientists want to predict when and where this breakdown occurs and to understand how organized atmospheric processes can give rise to smaller variations and turbulence.

Improved understanding of tropical-extratropical interactions. Atmospheric dynamics at midlatitudes play an important role in influencing the tropical atmosphere. The details of how midlatitude air motions interact with tropical air motions still need to be clarified. These details are particularly important for climate forecasting. In addition, the interactions among the Indian monsoon, the El Niño/Southern Oscillation, and other tropical processes must be better understood.

FUTURE DIRECTIONS IN CLIMATE RESEARCH

Theories concerning climate variability and climate change gained widespread interest in early 1970s and 1980s. The attention was spurred by weather-related disasters in many parts of the world, coupled with reports that human activities were altering concentrations of so-called greenhouse gases. Variations or changes in the earth's climate can influence many societal sectors, including agricultural yields, water availability and quality, transportation systems, ecosystems, and human health. In response to these issues, the U.S. National Climate Program and U.S. Global Change Research Program were initiated to "enhance and analyze observations, conduct process studies, and develop and improve climate models." The goal of these programs would be twofold. First, knowledge must be acquired concerning the physical, chemical, and ecological bases of climate. Such an understanding would allow scientists to characterize and predict the nature of climate variability from seasonal and yearly timescales to time periods of a decade and longer. Second, the relation of human activities and environmental resources must be completely assessed.

Over the past three decades, climate researchers have made considerable process in reaching the goals set out above. However, the many variables affecting climate and climate change, combined with the interactions of various physical, chemical, and other processes, mean that a number of climate research questions are still unanswered. Of all the subdisciplines of meteorology and the atmospheric sciences, climate research is probably the most ongoing. Some of the climate research areas likely to receive continued attention are as follows.

Understanding seasonal-to-yearly climate variability. For instance, how well understood are the driving processes of the El Niño/Southern Oscillation? To what extent can seasonal and year-to-year climate variations be predicted? A better understanding of processes at these timescales can improve predictions and knowledge of processes at larger timescales.

Understanding the nature of global and regional climate variability on seasonal-to-decadal and longer (for instance, decade to century) timescales. A significant problem in climate research involves separating the signal from the noise. Natural variability is a part of climate, but where does natural variability stop and some sort of human-induced or other signal begin? Current and future efforts will likely attempt to characterize the spatial and temporal aspects of this variability. In addition, increased examination will be focused on understanding, and eventually perhaps predicting, climate phenomena that acutely affect societies on a regional scale.

Understanding to what extent climate variations are predictable. Fluctuations in solar activity, sea surface temperature, and other naturally varying parameters can affect climate. In many cases, the effects of these fluctuations can be predicted, although the prediction accuracy may vary with time and location.

Identifying and obtaining the data needed to evaluate predictions. Quantitative assessments of prediction accuracy require high-quality measurements with which to compare. Identifying the times and locations at which monitoring can be useful is important to evaluating and improving the accuracy of climate predictions.

Understanding the earth's climate history and its causes. The earth's climate has seen enormous changes, from millennia of temperate and even tropical conditions to ice ages. Some understanding of the causes of these changes is essential to any study of global climate change.

Understanding the human-induced and natural changes to the global climate system. Natural climate variations are an innate part of the climate record. Do these variations explain any climate changes that have been observed? Do human-induced changes to the climate system provide a more likely explanation? These are important questions to which clear and indisputable answers must be found.

Understanding the response of climate system to changing greenhouse gas concentrations and water vapor and aerosol amounts. For instance, will a global warming due to increased greenhouse gases mean an increased frequency of El Niño occurrences? Any predictions or planning for climate change must rely on accurate understanding of these interrelations. However, in many cases, the climate system's exact responses can be difficult or even impossible to predict.

Understanding the extent to which climate can be simulated at an appropriate scale. Current simulations of climate change often do not provide the geographic grid spacing necessary to make quality estimates of regional effects. For instance, how might changes in climate affect agriculture or the spread of diseases? Improved understanding of these regional-scale simulations will help maximize their benefits to society.

Understanding the role of the stratosphere in surface and climate variability and anthropogenic change. Meteorologists are finding increasing evidence that an accurate representation of the stratosphere is critical to global climate models. Without this, the patterns of climate change cannot be fully understood. Improved computing resources will be important in attaining this goal, since addition of a stratosphere makes the models much more expensive computationally.

Defining a feasible and affordable climate-observing network. Climate observations have historically been based on the needs of operational weather forecasting and therefore have faced certain limitations. Climate observations are extremely valuable in providing any opportunities to detect climate changes. To be able to detect either climate stability or change requires stringent constraints for consistent and accurate observations. The role of satellite observations and the importance of ground-based observations need to be well defined to achieve a satisfactory climate-monitoring network. The free international exchange of climate data is another issue currently receiving discussion, though this issue leans more toward the political rather than the scientific.

Resolving model uncertainties due to clouds. Climate models have been used to establish a large base of knowledge about the behavior and interrelationship of various atmospheric constituents. The models are not able to account for all factors perfectly, however, and those areas remain the subject of emphasis for many climate researchers. One such area involves representations of the type and amount of cloud cover. The amount and vertical distribution of cloud significantly affects the earth's energy balance. Evidence has shown that clouds may contribute a positive feedback effect to climate warming. As the earth's surface temperature rises, leading to increased evaporation, cloud amount may also increase. Warmer temperatures in the lower atmosphere mean increased cloud formation at higher altitudes. These high-altitude clouds absorb and emit radiation in the atmosphere, thereby enhancing the greenhouse effect. An enhanced greenhouse effect would result in warmer surface temperatures, and the feedback loop continues. Obtaining more exact measurements of cloud amount and height, which can then be used to improve cloud parameterizations in climate models, is the objective of many field campaigns and satellite missions, discussed more extensively in Chapter 2.

Resolving model uncertainties due to aerosols. Aerosols are suspensions of particles in the atmosphere. The term *aerosol* can be considered to mean "dirty air." In the lower atmosphere, aerosols may consist of dust and pollutants from factories and car exhaust. These aerosols play an increasingly important role in the lower atmosphere. Countries reliant on significant coal use, including China and India, generate large amounts of sulfate aerosol in the lower atmosphere. Climate models have shown that the presence of such a layer may lead to a cooling effect at the surface, thus potentially offsetting the effects of global warming on a regional scale. Aerosols are also present in the stratosphere, as the end product of huge amounts of sulfur dioxide ejected into the atmosphere by volcanic eruptions. The nature and magnitude of aerosol effects on climate have been the subject of much recent research and will likely continue to remain in the forefront until a better understanding is reached.

Resolving other model uncertainties. The role of climate variability, land-surface processes, ocean circulation, and the interaction between chemistry and climate are other sources of uncertainty in climate models and subjects of current research. In an analysis of climate models, Eric Barron, professor and atmospheric scientist at Pennsylvania State University, concludes that "for many reasons—including the need for additional observational data—significant reductions in many of the uncertainties will require sustained efforts over a decade or more."

FUTURE DIRECTIONS IN BOUNDARY-LAYER METEOROLOGY RESEARCH

Boundary-layer meteorology relates to the layer of the atmosphere closest to the earth's surface. The boundary layer is in the lowest part of the troposphere. Its top may be as low as 10 meters (33 feet) or as high as 3 kilometers (1.9 miles). At the bottom of this layer, the atmosphere and surface meet, giving rise to turbulence. This is also the layer most directly heated by the terrestrial radiation emitted by the earth's surface. Meteorologists are often interested in boundary-layer processes and their relation to pollution transport. In addition, the boundary layer affects other processes, including cloud formation and transport of heat and water.

A number of goals related to current unresolved issues in boundary-layer meteorology research follow.

Attainment of increased understanding of the structure of cloudy boundary layers and the effects of boundary-layer clouds on climate. Water vapor present in the boundary layer, usually from evaporation from the surface, can form clouds. Boundary-layer clouds are usually one of two types: fair-weather cumulus cloud or a low-level stratocumulus cloud (low enough, the latter is called fog). Like most other boundary-layer processes, these clouds can be very short-lived, occurring on timescales of 1 hour or less. Nevertheless, boundary-layer clouds, particularly stratocumulus clouds over the oceans, can play a significant role in atmospheric processes and in climate.

Improvements in the current understanding of turbulence and boundary-layer mixing. Accurate modeling of pollution dispersion and other boundary-layer factors depends greatly on the knowledge available to prepare the model and perform the calculations. The small time and space scales associated with turbulence complicate both measurements and simulations of boundary-layer processes.

Improved measurements of water, heat, and other constituents at the earth's surface. The boundary layer plays an important role in the transport of heat, water vapor, dust, and smoke and other pollution. Better measurements of these quantities can improve understanding of how species are transported through the boundary layer and entire troposphere.

Improved understanding of boundary-layer interactions with the biosphere, including vegetation. The exchange of heat and moisture among land surface, atmosphere, and vegetation is an important factor in many processes involving air and water. Full knowledge of the interrelations between the atmosphere and biosphere is applicable to many fundamental climate issues, including improved understanding of the role of forest ecosystems in carbon dioxide exchange.

Attainment of increased understanding of boundary-layer, surface, and cloud interactions. Such information would be essential for improvements to models such as those currently used for daily temperature cycle estimates, hydrologic studies, and pollution prediction.

Use of new remote sensors to obtain more complete knowledge of the three-dimensional boundary-layer flow. The processes influencing winds and other atmospheric motions in the boundary layer are especially complex. New instruments, such as those described in Chapter 2, can provide a wealth of information to increase understanding of boundary-layer phenomena.

FUTURE DIRECTIONS IN ATMOSPHERIC RADIATION RESEARCH

The term *atmospheric radiation* is used to encompass all processes involving radiation from the sun and its interactions with the earth's atmosphere and surface. Because the amount of energy our planet receives from the sun is fixed (variations from sunspots and other solar activity are not particularly large), these processes can be summarized in a *radiation budget*—the incoming and outgoing radiation of the planet. Radiation is the driving force of all weather and climate on earth. Because of this, a thorough understanding of the processes governing the earth's radiation budget is a necessary component in many studies, from short-term weather prediction to long-term climate change.

The earth's atmosphere is composed of various gases and other constituents, including ice crystals and suspended small particles called aerosols. These gaseous and other species are responsible for absorption and scattering of solar radiation from the sun and terrestrial radiation from the earth's surface. Scientists have made significant progress in understanding how these gases and particles, as well as liquid water in clouds, interact with radiation. Among other topics, future research is likely to address three main issues, summarized below.

Improved understanding of the role of water vapor and cloud liquid water on absorption of solar and infrared radiation. Understanding how water vapor and cloud liquid water amounts affect the transfer of radiation through the atmosphere will increase scientists' knowledge of possible feedbacks to the climate system. Changes in water vapor or cloud liquid water could have important ramifications for climate.

Improved understanding of the role of nonspherical, irregular particles such as ice crystals and aerosols in scattering or reflecting radiation. Ice crystals and clouds can be efficient scatterers of solar radiation, thereby cooling the climate. The magnitude of this effect can vary greatly with the size and shape of the ice crystal or aerosol, so that overall climate impacts are not yet well-understood.

Improved understanding of the effect of clouds on the earth's radiation budget. This includes better estimates of the spatial distributions and radiative properties of various cloud types, from low-level stratus clouds often present over the oceans to more scattered upper-level cirrus. This information will have value in many areas of meteorological research. Two areas for which the information is essential concern the improvement of radiative transfer models and improvements of cloud and radiative transfer initializations and calculations in general circulation models.

FUTURE DIRECTIONS IN CLOUD PHYSICS RESEARCH

Clouds play a major role in the earth's radiation budget by scattering and absorbing solar and terrestrial radiation. For years, scientists have studied the formation processes, distributions, and compositions of various types of clouds. This work has relied on surface observations of cloud type and extent, satellite observations of the frequencies of occurrence and spatial distributions of various cloud types, and other remote sensing measurements providing information on cloud composition (see Chapter 2 for a definition of *radar* and *lidar*). In addition, in situ observations from balloons or aircraft, which can carry instruments directly into the cloud, or studies in laboratory simulators called cloud chambers also provide information on clouds and cloud composition.

Future work in cloud physics is likely to focus on a number of related and clearly defined issues. A list of these research goals follows.

Prediction of the extent, lifetime, composition, and radiative properties of often-occurring mid-level stratocumulus and upper-level cirrus clouds. Such information is necessary in studies of atmospheric-ocean interactions and influences on the earth's water budget. This knowledge would also improve understanding of the effects of natural and pollutant aerosols in cloud reflectivity and the role of cirrus clouds in climate change.

Advancement of the understanding of the interactions of clouds and radiation. Outstanding issues in this area concern: the ability to estimate the coverage and radiative properties of various cloud types; a better understanding of liquid and solid (snow, ice) precipitation formation; precipitation mechanism theories such as warm-rain and ice-phase precipitation processes; the effects of precipitation production and evaporation on storms and storm evolution; and the influence of precipitation processes on advertent and inadvertent weather modification (see Chapter 3).

Prediction of atmospheric aerosol and water droplet size ranges to aid in determining: their joint influence on the radiation budget; their role in chemical reactions and precipitation formation; and their influences on cloud composition and climate effects.

Improvement of the understanding of interactions among aerosols, chemical species, and clouds. This information will be particularly useful for the improvement of global chemical models.

FUTURE DIRECTIONS IN ATMOSPHERIC ELECTRICITY RESEARCH

The study of atmospheric electricity represents a specialized area of emphasis within atmospheric physics. The most common example of atmospheric electricity is the cloud-to-ground lightning often observed during thunderstorms. Most of this lightning is generated in the lower part of the cloud and delivers a negative electric charge to the ground. However, electrical phenomena also occur at higher altitudes and may be associated with lightning that has a positive charge. The field continues to present many mysteries, and scientists working on atmospheric electricity issues are likely to direct future research toward the following objectives.

Determination of mechanisms for charge generation and separation in clouds. This knowledge will assist in understanding cloud formation processes and improve understanding of the fundamental physics of lightning.

Determination of the nature and sources of middle-atmosphere electric discharges. There is little understanding of the possible association between middle-atmosphere discharges and severe weather. These discharges are known to affect how radio waves propagate and can also influence atmospheric chemistry.

Understanding the role of mid- and upper-atmosphere electrical phenomena in the planetary energy budget. Recently discovered atmospheric electrical phenomena include split-second flashes of colored light that occur above thunderstorms at altitudes up to almost 100 kilometers (60 miles). These electrical phenomena are called sprites and elves and occur in a part of the atmosphere long thought to be inert electrically. Their existence likely plays a role in the exchange of energy between the earth and space.

Quantification of the production of nitrogen oxides by lightning. Nitrogen oxides have been found to play an important role in ozone production and destruction. Quantification of lightning-produced nitrogen oxide amounts will provide a better understanding of their correlation to ozone production and loss in the upper troposphere.

Investigation of the possibility that changes in global and regional lightning frequencies might be an indicator of climate change. Recent climate records document changes in surface temperatures, precipitation, and other climate variables. Lightning is an additional variable in which changes may become evident.

FUTURE DIRECTIONS IN ATMOSPHERIC CHEMISTRY RESEARCH

Atmospheric chemistry involves the study of "environmentally important atmospheric species" (NRC, 1998). These are species that, by virtue of their chemical or radiative properties, affect climate, ecosystems, and humans and other organisms. Although chemistry as a discipline has been around since the eighteenth century, it was not until the latter part of the twentieth century that atmospheric chemistry emerged as a quantitative, scientific discipline all its own.

Chapter 1 contains extensive discussions of current atmospheric chemistry issues, including tropospheric pollution, acid rain, and the stratospheric ozone hole. Given the science's relative youth, it is sometimes astonishing to consider the wealth of new information. Future research in atmospheric chemistry is likely to be aimed toward the following goals.

Documentation of the variability and long-term trends associated with many chemical species in the atmosphere. Understanding the spatial and temporal distributions of various chemical species is essential to many topics in meteorological and climate research, particularly as related to issues of human-induced climate change. The proposed documentation can be accomplished only through development and maintenance of high-quality monitoring networks to obtain high-quality regional and global-scale observations.

Development of predictive tools and models by synthesizing information from field studies, laboratory experiments, and other observational efforts. Prior to thirty years ago, very few people worked in or knew much at all about the atmospheric chemistry discipline. Only in recent decades have atmospheric chemistry issues gained international attention. In recent years, full-fledged research campaigns have been devoted to gathering new knowledge about many of the atmosphere's chemical constituents. The synthesis of much of this work by meteorologists, atmospheric scientists, and chemists in the field and in the lab will dramatically increase understanding of chemical species and processes.

Representation of chemistry in mathematical and numerical models. Accurate representation of chemical processes in global and regional climate models will do much to improve the model predictions as well as scientists' understanding of the interactions of various chemical and physical processes. Current and future research efforts in this area require the collaboration of scientists from a variety of disciplines, including chemistry, meteorology and other geophysical sciences, and mathematics.

A holistic and integrated study of environmentally important atmospheric species and the chemical, physical, and ecological interactions that couple them together. A thorough understanding of the roles of various atmospheric species can be obtained only through knowledge of the chemical, physical, and ecological processes involved. The interactions among these three processes must also be better understood for any environmentally important atmospheric species assessment to be complete.

Documentation of the distributions, variability, and trends in stratospheric ozone. This also includes documentation of the distribution and trends in species responsible for ozone destruction, including chlorofluorocarbons and methyl chloroform. Studies of ozone and other species distribution will also provide information relating chemistry, dynamics, and radiation in the stratosphere and upper troposphere.

Improvement of scientists' understanding of processes that control the abundance, variability, and long-term trends of carbon dioxide, methane, nitrous oxide, ozone, and water vapor. These species are especially important to global climate studies. Understanding of these species can be improved by expanding global monitoring programs to include measurements of upper-tropospheric and lower-stratospheric ozone and water vapor.

Development of observations and computational tools to manage tropospheric ozone pollution. This process requires understanding the interrelations among species responsible for tropospheric ozone formation to control and predict poor air quality levels.

Documentation of the physical, chemical, and radiative properties of atmospheric aerosols, including their spatial extent and long-term trends. Aerosols play an important role in climate by scattering atmospheric radiation. Full understanding of these effects requires improved knowledge of the physical and chemical processes governing size ranges and number concentrations.

Documentation of the rates of chemical exchange between the atmosphere and key ecosystems. Chemical species in the atmosphere can act as either toxics or nutrients to various ecosystems. By understanding the influence of changing concentrations and the deposition of harmful or beneficial compounds on the atmosphere and biosphere, scientists can make recommendations for better protecting ecosystems that may be of economic or environmental importance.

Other specific issues relating to atmospheric chemistry include:

Estimation of Mexican emissions and their effects on air quality in the United States. Although it is known that emissions in Mexico contribute significantly to visibility impairment in the United States, the nature and amount of these sources are not well documented. Control strategies in the United States could be offset by increased emissions across the border, making field and modeling studies of Mexican emissions essential.

Improved air quality monitoring methodologies, particularly in the areas of visibility impairment and optical absorption. High-quality, routine field observations are necessary to obtain a better understanding of natural conditions that limit visibility and of impairments due to sulfate, carbon-based, and other particles. In recent years, budget shortfalls have caused the discontinuation of monitoring efforts in many U.S. national parks. Continuation or expansion of current monitoring efforts is essential to understanding the effects of urban and other growth.

TOOLS FOR ACHIEVING THIS FUTURE

A number of technology improvements have been proposed to address these topics and are discussed in greater detail in Chapter 2. In addition, researchers are working on a number of new analysis techniques, including the development of multispectral algorithms to infer optical depth, cloud liquid water, and trace gas concentrations and the use of pattern recognition, artificial intelligence, chaos theory, and computer visualization techniques. Although descriptions of these methods are beyond the scope of this text, it is clear that future advances in meteorology will likely benefit greatly from new developments in these areas.

References
Barron, E.J. "Climate Models: How Reliable Are Their Predictions?" *Consequences*, 1(3): 17–27, 1995.
National Aeronautics and Space Administration, <http://www.nasa.gov>.
National Oceanic and Atmospheric Administration, <http://www.noaa.gov>.
National Research Council. *The Atmospheric Sciences Entering the Twenty-first Century*. Washington, DC: National Academy Press, 1998.
Robinson, W. Personal communication, 2000.

Chapter Five

Biographical Sketches

The men and women who work on issues in meteorology and the atmospheric sciences come from a variety of backgrounds. Although a physical science background is most common, study in other areas can also be applied to issues in meteorology, particularly in the areas of policy and societal effects. As in other fields of science, progress in meteorology is typically the result of many scientists working together.

Atmospheric research is usually the realm of persons holding a doctorate, a pattern that will be evidenced by the biographies below. Research teams also involve persons holding master's degrees and may include graduate students as well. Because of the international scope of many atmospheric issues, researchers from several countries often collaborate on a single project. The biographical sketches that follow typically profile persons leading a research team. Each of the individuals listed contributed directly to advancing our understanding of meteorology and atmospheric science. They have usually been widely recognized for their work and in many cases have received awards for outstanding performance or contributions to the science.

Howard Bluestein (1948–)

Tornado researcher Howard Bluestein has spent over twenty years observing and studying weather phenomena on convective, mesoscale, and synoptic scales. Bluestein's interest in weather phenomena, and particularly in violent storms, began when he was growing up near Boston, Massachusetts. He received a bachelor's degree in electrical engineering, master's degrees in electrical engineering and meteorology, and a Ph.D. in meteorology from the Massachusetts Institute of Technology. Bluestein is currently professor of meteorology at the University of Oklahoma in Norman and has often been a visiting scientist at the National Center for Atmospheric Research in Boulder, Colorado. He has used visual observations and portable Doppler radar to probe the flow pattern in tornadoes during intensive field studies. He is interested in the convective structure of the hurricane eyewall and has flown into the eye of hurricanes six times. The results of his research have been published in leading scientific journals, and his cloud photographs have appeared worldwide in numerous publications and in museums. Bluestein is a fellow of the American Meteorological Society. He has authored a textbook on synoptic-dynamic meteorology and another book on tornadoes of the U.S. plains.

Harold Brooks (1959–)

Harold Brooks is best known for his work on issues related to numerical weather prediction, forecast verification, modeling of hazardous weather, and other topics. He received a B.A. in physics and mathematics from William Jewell College in his home state of Missouri, an M.A. in atmospheric sciences from Columbia University, and a Ph.D. in atmospheric sciences from the University of Illinois. After completing his Ph.D., Brooks accepted a National Research Council research associateship at the National Severe Storms Laboratory in Norman, Oklahoma, where he has been since 1990. He currently serves as head of the Mesoscale Applications Group and has twice received the Environmental Research Laboratories' Outstanding Paper Award.

Stanley Changnon (1928–)

Stanley Changnon has studied climate variability on scales from decades to centuries and has worked extensively to relate climate changes to agriculture, water resources, and government. Changnon holds an M.S. from the University of Illinois and is an emeritus professor of geography and emeritus chief of the Illinois State Water Survey. His research focuses on physical climatology, hydrometeorology, hydrology, and agriculture. Changnon is a fellow of the Illinois

State Academy of Science and has published prolifically on climate topics and their relation to society.

Jule Charney (1917–1981)

Jule Charney is one of many earlier contributors to meteorology, and his work in numerical weather prediction has been incredibly long lasting. Charney headed the meteorology group responsible for the first computer-generated weather forecast at the Institute for Advanced Study at Princeton University. He was a pioneer in the study and modeling of atmospheric motions and a teacher to many of the first advanced-degree recipients in the field of meteorology. Charney studied mathematics and physics as an undergraduate at the University of California at Los Angeles (UCLA). He received a master's degree in mathematics and a Ph.D. in meteorology, both from UCLA. He began work at Princeton in 1948 and oversaw the first numerical weather prediction run in April 1950. Charney accepted a faculty position at the Massachusetts Institute of Technology in 1956. He served on many national panels and on the President's Science Advisory Committee and was an active contributor to the field through 1978.

John Christy (1951–)

John Christy conducts research related to climate dynamics, global-scale processes, and satellite temperature data and in 1996 received a special award from the American Meteorological Society "for developing a global, precise record of earth's temperature from operational polar-orbiting satellites." Christy holds a Ph.D. in atmospheric sciences from the University of Illinois and is professor and director of the Earth System Science Laboratory at the University of Alabama at Huntsville. He has received the NASA Medal for Exceptional Scientific Achievement and is a contributor to the Intergovernmental Panel on Climate Change and National Research Council's Space Studies Board and Panel on Atmospheric Temperature.

Ralph Cicerone (1944–)

Ralph Cicerone has contributed enormously to research improving scientists' understanding of the atmosphere-earth system and was one of six American scientists to receive a United Nations award for research to protect the earth's ozone layer. He holds a bachelor's degree in electrical engineering from the Massachusetts Institute of Technology and an M.S. and Ph.D. from the University of Illinois. Cicerone held positions as an assistant professor in the Department of Electrical and Computer Engineering at the University of Michigan and as a research chemist at Scripps Institution of Oceanography. He served as a senior scientist and director of the Atmospheric Chemistry Division at the National Center for Atmospheric Research. In 1989, he joined the Earth System Science Department at the University of California at Irvine and was appointed chancellor of the university in April 1998.

Paul Crutzen (1933–)

In 1995, Paul Crutzen, along with Mario Molina and Sherwood Rowland, received the Nobel Prize for chemistry for work on the chemical reactions leading to ozone depletion in the stratosphere. Trained as a civil engineer in his native Amsterdam, Crutzen earned a D.Sc. and Ph.D. in meteorology from the University of Stockholm. In 1974, he traveled to the United States to join the National Center for Atmospheric Research in Boulder, Colorado, where he was eventually appointed senior scientist and director of the Air Quality Division. During this time, he was also a consultant to the National Oceanic and Atmospheric Administration's Aeronomy Laboratory and an adjunct professor in the atmospheric sciences department at Colorado State University. In 1980, Crutzen returned to Europe to take a position at the Max Planck Institute for Chemistry in Mainz, Germany. Even while spending most of his time in Europe, Crutzen has served as a part-time professor in the Department of Geophysical Sciences at the University of Chicago and at the Scripps Institute of Oceanography at the University of California at San Diego. He has received numerous honors and awards for his work in atmospheric chemistry and other issues.

Charles Doswell III (1945–)

Throughout his career, Charles Doswell has devoted extensive study to the field of weather prediction and has conducted numerous experiments to improve weather forecasting. Doswell holds a B.S. in meteorology from the University of Wisconsin in Madison and an M.S. and Ph.D., also in meteorology, from the University of Oklahoma. He began his professional career in 1976 at the National

Severe Storms Forecast Center in Kansas City, Missouri, doing both research and severe weather forecasting. In 1982, he moved to the Environmental Research Laboratories in Boulder, Colorado, where he worked with the Weather Research Program. He moved to the National Severe Storms Laboratory in 1986, where he holds a position as a research meteorologist. Doswell has authored or coauthored more than 150 scientific papers. He has received numerous National Oceanic and Atmospheric Administration Sustained Superior Performance awards, including a Department of Commerce Silver Medal Award. Doswell is an adjunct faculty member with the University of Oklahoma's School of Meteorology and occasionally teaches various courses, including his own graduate-level course in advanced forecasting techniques.

Paul Douglas (1958–)

Paul Douglas has been chief meteorologist for KARE-11 News in Minneapolis–Saint Paul since 1983. He is also author of *Prairie Skies: The Minnesota Weather Book*. At the request of a radio station owner who employed him early in his career, Douglas Paul Kruhoeffer simplified his name for radio and TV. He received a degree in meteorology from Penn State, and worked at television stations in Pennsylvania and Connecticut before moving to Minnesota. Douglas has served as a member of the Broadcast Board of the American Meteorological Society and provides forecasts and a daily weather column for the *Minneapolis Star Tribune*. He has also served as president of Total Weather, which prepares detailed forecasts for weather-sensitive companies.

Michael Glantz (1939–)

Michael Glantz studies how climate affects society and how society affects climate, and he is an expert in the interaction between climate variation and human activities and how these changes affect quality-of-life issues. Glantz received his B.S. in metallurgical engineering, and M.A. and Ph.D. in political science from the University of Pennsylvania. He is a senior scientist in the Environmental and Societal Impacts Group at the National Center for Atmospheric Research and was the program director of the group from 1979 to 1997. He has edited over seventeen books, authored over 100 research articles, and is a member of numerous national and international committees and advisory bodies related to environmental issues. In 1990 he received the prestigious Global 500 award from the United Nations Environment Programme.

James Hurrell (1962–)

James Hurrell is well known for his work on climate change and variability, which he studies using data collected by satellite instruments and other observing systems. He holds a B.S. in mathematics and earth/space science from the University of Indianapolis and an M.S. and Ph.D. in atmospheric science from Purdue University. He is a scientist in the Climate Analysis Section at the National Center for Atmospheric Research. In 1997, Hurrell received the National Center for Atmospheric Research's Outstanding Publication Award.

Bryan Johnson (1958–)

Bryan Johnson has worked extensively on ozone measurements and data analysis, and currently coordinates ozonesonde projects for the Climate Monitoring and Diagnostics Laboratory (CMDL) of the National Oceanic and Atmospheric Administration (NOAA). He received a bachelor's degree in chemical engineering from Montana State University, an M.S. degree in meteorology from South Dakota School of Mines and Technology, and a Ph.D. in atmospheric sciences from the University of Arizona. After receiving his Ph.D., Johnson accepted a postdoctoral research associate position with the Department of Atmospheric Sciences at the University of Wyoming. In 1994, he joined the Cooperative Institute for Research in Environmental Sciences at the University of Colorado and NOAA and is currently a research chemist in the Ozone and Water Vapor Group at NOAA/CMDL.

Eugenia Kalnay (1942–)

Eugenia Kalnay has been a pioneer in applying ensemble forecasting techniques and has published extensively on issues related to numerical weather prediction and atmospheric dynamics. Kalnay is professor and chair of the Department of Meteorology at the University of Maryland. She is a former Robert E. Lowry Chair in Meteorology at the University of Oklahoma and former director of the Environmental Modeling Center at the National Centers for Environmental Prediction. Kalnay holds a license in meteorology from the University of

Buenos Aires and a Ph.D. from the Massachusetts Institute of Technology. She has received the American Meteorological Society's Jule G. Charney Award and a NASA Medal for Exceptional Scientific Achievement, as well as several Department of Commerce Gold Medals, and a Senior Executive Service Presidential Rank Award.

Thomas Karl (1951–)

Thomas Karl is internationally recognized for his work on climate and change and serves as co-chair of the U.S. National Climate Assessment. Since 1990, he has been a lead author on each of the assessments of the Intergovernmental Panel of Climate Change. Karl holds a master's degree in meteorology from the University of Wisconsin. He serves as the director of the National Climatic Data Center of the National Oceanic and Atmospheric Administration (NOAA) and also manages the Climate Change Data and Detection Program Element for NOAA's Office of Global Programs. Karl is a fellow of both the American Meteorological Society and the American Geophysical Union and recently completed chairing the National Academy of Sciences Climate Research Committee. Karl has received numerous awards, including the Helmut Landsberg Award, the Climate Institute's Outstanding Scientific Achievements Award, two Department of Commerce Gold Medals, one Bronze Medal, and the NOAA Administrator's Award. Karl has authored nearly 100 peer-reviewed journal articles, been coauthor or coeditor on numerous texts, and has published over 200 technical reports and atlases.

Charles David Keeling (1928–)

Charles Keeling played an important role in measuring atmospheric carbon dioxide at the Mauna Loa Observatory, work that earned him the 1980 Half Century Award of the American Meteorological Society. Keeling holds a B.A. in chemistry from the University of Illinois and a Ph.D. in chemistry from Northwestern University. He is a professor in the geosciences research division at the Scripps Institution of Oceanography, where his interests include the geochemistry of carbon and oxygen, emphasizing the carbon cycle in nature and the abundance and air sea exchange of carbon dioxide. Keeling has served as the scientific director of the Central CO_2 Laboratory of the World Meteorological Organization and has been a Guggenheim fellow at the Meteorological Institute, University of Stockholm, Sweden. He is the author of nearly 100 research articles and has served on several scientific committees and panels.

Michael McPhaden (1950–)

Michael McPhaden is well known for his work on La Niña and its effects on climate. He received a bachelor's degree in physics from the State University of New York at Buffalo and a Ph.D. in physical oceanography from Scripps Institution of Oceanography. Following his Ph.D., McPhaden spent time as a research scientist at the National Center for Atmospheric Research, then moved to the University of Washington. He is currently a senior research scientist at the Pacific Marine Environmental Laboratory in Seattle, Washington, and also serves as an affiliate professor and senior fellow of the Joint Institute for the Study of the Atmosphere and Ocean. He has received numerous honors and awards, including the Department of Commerce Gold Medal, several Outstanding Performance and Outstanding Publication awards of the National Oceanic and Atmospheric Administration, and was the Sverdrup Lecturer at the American Geophysical Union fall 1998 meeting.

Mario Molina (1943–)

Mario Molina has contributed greatly to work documenting the effects of chloroflurocarbons on the stratosphere and in 1995 was a corecipient of the Nobel Prize for chemistry. Molina was educated internationally, attending private school in Switzerland and completing his precollege education in his native Mexico. He studied in Germany and Paris before enrolling in the Physical Chemistry Department at the University of California at Berkeley. After completing his Ph.D. at Berkeley, he began a postdoctoral appointment at the University of California at Irvine. Under the direction of Sherwood Rowland, he set to work understanding the environmental fate of chlorofluorocarbons. After nine years at Irvine, Molina left in 1982 to take a research scientist position at the Jet Propulsion Laboratory in Pasadena, California. Six years later, he accepted a joint appointment as professor in the Departments of Chemistry and of Earth, Atmospheric, and Planetary Sciences at the Massachusetts Institute of Technology. He is currently MIT's Martin Professor of Atmospheric Chemistry.

Stephen Montzka (1961–)

Stephen Montzka has made important contributions in measuring the amount of ozone-depleting chemicals in the atmosphere. Montzka received a bachelor's degree in chemistry from Saint Lawrence University. After completing a Ph.D. in analytical chemistry at the University of Colorado in Boulder, he held a two-year position as a National Research Council postdoctoral fellow in the National Oceanic and Atmospheric Administration's (NOAA) Aeronomy Laboratory. He has been a research chemist in NOAA's Climate Monitoring and Diagnostics Laboratory since 1991. In 2000, he was selected as a NOAA Employee of the Year.

James O'Brien (1935–)

James O'Brien conducts research directed principally toward understanding upper ocean dynamics on timescales of several days to several years. He has worked on developing models of the equatorial circulation, upper ocean, and ice-ocean interaction. These models require supercomputers to run and provide very good descriptions of the processes associated with El Niño. O'Brien holds a Ph.D. from Texas A&M University. He is a professor of meteorology and oceanography at Florida State University. He and his students have investigated how El Niño and La Niña directly affect hurricanes, tornadoes, snowfall, forest fires, and other climate phenomena across the United States.

Roger Pielke, Jr. (1968–)

Roger Pielke, Jr. explores how scientific information relates to public and private sector decision making. His current topics of interest include societal responses to extreme weather events, domestic and international policy responses to climate change, and science policy in the United States. Pielke holds a B.A. in mathematics and a Ph.D. in political science from the University of Colorado. He is a scientist in the Environmental and Societal Impacts Group at the National Center for Atmospheric Research and chairs the American Meteorological Society's Committee on Societal Impacts. In 2000, he received the Sigma Xi Distinguished Lectureship Award. He serves on the Science Steering Committee of the World Meteorological Organization's World Weather Research Programme, as well as on other advisory committees. Pielke is coauthor or coeditor of three books, most recently *Prediction: Decision Making and the Future of Nature*, with D. Sarewitz and R. Byerly.

Prabhakara Cuddapah (1934–)

C. Prabhakara conducts research related to climate analysis and utilization of remote sensing data and has received two awards for exceptional performance as well as recognition for exceptional achievement. Prabhakara received a Ph.D. in meteorology from New York University and is a scientist at NASA's Goddard Space Flight Center. He is a member of the American Meteorological Society's committee on climate variability. His most recent work has focused on deriving and analyzing tropospheric temperatures from satellite data.

F. Sherwood Rowland (1927–)

Sherwood Rowland shared the 1995 Nobel Prize for chemistry with Mario Molina and Paul Crutzen. Rowland contributed to important findings concerning the chemical reactions of ozone depletion in the stratosphere and specifically the effects of chlorofluorocarbons. After receiving a Ph.D. in chemistry from the University of Chicago, Rowland taught chemistry at Princeton University and then at the University of Kansas. In 1964, he was appointed professor of chemistry and department chair at the new University of California campus at Irvine. It was following retirement from this position in 1970 that Rowland became interested in the environmental chemistry of chlorofluorocarbons (CFCs). Rowland and his postdoctoral fellow, Mario Molina, were able to show how CFCs can break apart in the stratosphere, and play a role in destroying ozone.

Stephen Schneider (1945–)

Stephen Schneider is one of the world's experts on the implications of climate change for the environment and society. Schneider received a Ph.D in plasma physics from Columbia University. He spent more than two decades as a climate researcher at the National Center for Atmospheric Research, and is currently a professor in the Biological Sciences Department at Stanford University. While at NCAR, he served as head of the Advanced Study Program. He has authored numerous books, most recently *Laboratory Earth* (1998).

Susan Solomon (1956–)

Susan Solomon is a specialist in upper-atmosphere chemistry and in 2000 received the President's Science Medal for her contributions verifying the effects of chlorofluorocarbons in the polar stratosphere. Solomon completed her undergraduate degree in chemistry at the Illinois Institute of Technology in Chicago and went on to receive a doctoral degree from the University of California at Berkeley. Following her Ph.D. work, she accepted a position with the National Oceanic and Atmospheric Administration's Aeronomy Laboratory in Boulder, Colorado. Solomon led a group of researchers to Antarctica in 1986 to measure gases in the polar stratosphere. Their results verified the role of polar stratospheric clouds in chemical reactions and provided important information on polar ozone chemistry. Since that time, she has worked on improving the understanding of how atmospheric chemistry affects climate.

Richard Stolarski (1941–)

Richard Stolarksi has contributed greatly to scientists' understanding of ozone chemistry and has been awarded the U.S. Environmental Protection Agency Stratospheric Ozone Protection Award, the United Nations Environmental Programme Global Ozone Award, and a NASA Exceptional Achievement Award. Stolarski holds a B.S. in physics and mathematics from the University of Puget Sound and a Ph.D. in physics from the University of Florida. Stolarski held positions at the University of Michigan and at NASA's Johnson Space Center before joining NASA's Goddard Space Flight Center in 1976. From 1979 to 1985, he was head of the Atmospheric Chemistry and Dynamics branch. Stolarski is a fellow of the American Geophysical Union.

Gary Thomas (1934–)

Gary Thomas specializes in research in radiative transfer and the photochemistry of the earth's atmosphere. His current research topics include the structure and dynamics of the earth's upper atmosphere, noctilucent clouds, and middle-atmospheric climate change. Thomas uses several techniques to accomplish these objectives, including theoretical modeling of satellite and rocket data. He holds a Ph.D. from the University of Pittsburgh and is a professor of astrophysical and planetary sciences at the University of Colorado.

Kevin Trenberth (1944–)

Kevin Trenberth is an internationally recognized authority on ocean-atmosphere interactions and currently heads the Climate Analysis Section at the National Center for Atmospheric Research (NCAR). He received his Sc.D. from the Massachusetts Institute of Technology and worked as a meteorologist in his home country of New Zealand before spending time as a faculty member at the University of Illinois. He has been with NCAR since 1984. Trenberth is a fellow of the American Meteorological Society and the American Association for the Advancement of Science and an honorary fellow of the Royal Meteorological Society of New Zealand. He is the recipient of the American Meteorological Society's Jule G. Charney Award and has written or contributed to twenty-three books and book chapters and over 100 refereed journal articles.

Hans von Storch (1949–)

Hans von Storch is known for his work in the statistical analysis of geophysical data, simulation of regional climates, and paleoclimate modeling. He received a diploma in mathematics and a Ph.D. in meteorology from the University of Hamburg. From 1987 to 1995, he was a senior scientist and leader of the Statistical Analysis and Modeling Group at the Max Planck Institute for Meteorology. von Storch is director of the Institute of Hydrophysics of the GKSS Research Centre in Geesthacht and a professor at the Meteorological Institute of the University of Hamburg. He has published six books and numerous scientific articles and is lead author of the regional climate chapter of the Third Assessment Report of the Intergovernmental Panel on Climate Change.

Elizabeth Weatherhead (1958–)

Elizabeth Weatherhead is an expert in ultraviolet (UV) radiation monitoring and in detection of trends. Her work has focused on UV radiation monitoring and on detection of environmental change, including issues concerning detection of trends in atmospheric ozone and other climate variables. Weatherhead holds a bachelor's degree in physics, master's degrees in physics and statistics, and a doctoral degree in geophysical sciences, all from the University of Chicago. After completing her Ph.D., she was a visiting scientist at the Non-Ionizing Radiation Laboratory in Stockholm, Sweden, and at the

World Health Organization's International Agency for Research on Cancer in Lyon, France. In 1993, she accepted an appointment with the Cooperative Institute for Research in Environmental Sciences at the University of Colorado and the Air Resources Laboratory of the National Oceanic and Atmospheric Administration.

Joshua Wurman (1960–)

Joshua Wurman has been a leader in developing the Doppler on Wheels (DOW) mobile weather radars for tornado and other severe storms research. His team successfully measured wind speeds in the 1999 Oklahoma City tornado and has collected unprecedented data revealing the structure, formation processes, and evolution of tornadoes. Wurman received a doctoral degree from the Massachusetts Institute of Technology and is currently associate professor of meteorology at the University of Oklahoma in Norman. His development work on Doppler radar networks has been done in collaboration with McGill University in Montreal, Canada, and the National Center for Atmospheric Research in Boulder, Colorado. Wurman's main research interests are in mesoscale meteorology, meteorological radar development, and instrumentation.

Edward Zipser (1937–)

Edward Zipser has contributed significantly to the understanding of weather processes, including the concept of unsaturated mesoscale downdrafts. He has led many coordinated field programs and aircraft missions to investigate various weather phenomena. Zipser received a bachelor's degree in aeronautical engineering from Princeton University and an M.S. and Ph.D. in meteorology from Florida State University. From 1966 to 1990, Zipser was a scientist with the National Center for Atmospheric Research in Boulder, Colorado. In 1990, he moved to Texas A&M University to teach and head the Department of Meteorology. In 1999, he left Texas A&M for his current position as professor and chair of the Department of Meteorology at the University of Utah. Zipser was named a fellow of the American Meteorological Society in 1982 and has received numerous awards for outstanding contributions and leadership.

Chapter Six

Career Information

The American Meteorological Society defines a *meteorologist* as "a person with specialized education who uses scientific principles to explain, understand, observe or forecast the earth's atmospheric phenomena and/or how the atmosphere affects the earth and life on the planet." A meteorologist is essentially a scientist of the atmosphere. *Atmospheric science* is a broader term, usually referring to the combination of meteorology with other branches of physical science, such as chemistry or physics. People with advanced degrees in atmospheric science or meteorology work in research, teaching, forecasting, or private industry.

Currently, there are about 20,000 meteorologists and atmospheric scientists in the United States. These people typically hold at least a bachelor's and often a master's or doctoral degree in meteorology or atmospheric science, or related fields, including physics, chemistry, mathematics, or earth system science. Career opportunities in meteorology include forecasting, research, teaching, broadcasting, and consulting. This chapter describes the major meteorology-related occupations and provides information on the educational and training requirements associated with each.

EMPLOYMENT SECTORS

Operational Meteorology

Operational meteorologists devote their talents to generating short-term and long-term forecasts for use by the general public or by weather-affected industries. Accurate information regarding current and upcoming weather conditions, especially in terms of severe or extreme events, is important for safety and economic reasons. The meteorologists responsible for these forecasts use measurements of temperature, air pressure, humidity, and wind velocity at various levels of the atmosphere. The data used are referred to as *synoptic data*, describing large-scale atmospheric motions, and the scientists using them are sometimes referred to as *synoptic meteorologists*. The data come from balloon measurements, satellite observations, weather radar, and observers in many parts of the world and are input to sophisticated computer models to predict conditions for a given area at a given time.

Operational meteorology positions often involve rotating work schedules, including night and weekend shifts, since most forecast offices operate around the clock. The work environment is typically casual, and meteorologists may work alone or in teams, depending on the size of the office. People in these positions, as in other areas of meteorology, do their work using computers, maps, and charts of numeric data.

Traditionally, the most significant employer of operational meteorologists has been the National Weather Service, although recent restructuring within this organization has reduced the number of positions available. Also, contrary to what many may think, one does not simply complete a bachelor's degree and apply for a job as forecaster. Nearly all meteorologists with a degree start at the National Weather Service as an intern. Many of these interns were co-op students while they were still in school. After about three to six years and completion of extensive in-house training, an intern can advance to the level of journeyman forecaster, and from here to senior forecaster. A forecaster can then advance to a position as a warning coordination meteorologist or a science and operations officer. Each change of level requires an additional accrual of experience and training, and to facilitate this, the National Weather Service offers forecasting courses in-house and at specialized training centers.

The late 1990s brought an increase in the demand for meteorologists in the private sector. As more and more data products become available, these meteorologists are charged with interpreting the information and providing specialized forecasts for select clients, such as utility, aviation, shipping, fishing, or agricultural industries. Private sector meteorology careers are examined in greater detail below.

Figure 6.1. Meteorologists analyzing weather data displayed by the Advanced Weather Interactive Processing System (AWIPS). Courtesy of Wilfred von Dauster, National Oceanic and Atmospheric Administration (NOAA).

Broadcast Meteorology

Broadcast meteorologists are responsible for the communication of forecast information to the public, usually by means of television and radio. In his book *Prairie Skies*, television meteorologist Paul Douglas offers an inside look at the field and briefly examines the history of broadcast meteorology.

Broadcast meteorology is a subdiscipline that first gained visibility in the 1960s when television stations began hiring professional newscasters, usually middle-aged to older men, to handle weather segments on the local news. In the 1970s, communication of weather information was increasingly handled by female newscasters, unflatteringly known as "weather girls," and by comedians and other nonmeteorologists. Exceptions to this rule were areas where weather was of large interest or importance. In these areas, station directors began hiring meteorologists to interpret weather information and accurately communicate this to the audience. Still, by the 1990s, only about 30 percent of weathercasters on television or radio had had any formal meteorological education. These people have traditionally been employed mostly in the Midwest, where tornadoes and blizzards are serious and not uncommon threats. Minneapolis–St. Paul, Tulsa, Oklahoma City, Boston, Denver, and Kansas City are among the cities where most weathercasts are delivered by degreed meteorologists.

Broadcast meteorologists start their days early or work very late. Those preparing forecasts for the morning news report, for example, are usually at the station by 3:00 or 3:30 A.M. Their shift ends at around noon, and the "evening" person comes in at 2:00 P.M. and works until about 11:00 P.M. If severe weather is a threat, often a person will end up staying until midnight or later. Broadcast meteorologists do most of their work in a room filled with computers; the room may contain a camera for live weather broadcasts. And in broadcast meteorology as well as in the National Weather Service, there must be people to cover weekends and holidays.

Some strictly behind-the-camera television meteorology jobs do exist. These people typically work as weather producers, preparing the graphics and other information for the person who delivers the forecast. Aspiring broadcast meteorologists may initially work as weather producers as a way to gain experience and prepare for a role on camera, although this is not always the case. The meteorological expertise provided by a weather producer is an essential part of a successful weathercast, especially when the broadcast is delivered by a nonmeteorologist. Employment as a weather producer requires a good understanding of weather processes and computer graphics and can be a way for camera-shy meteorologists to contribute to the communication of weather information.

Research Careers

Some scientists focus investigations on the physical and chemical properties of the atmosphere, studying such things as the formation of clouds, rain, snow, storms and other weather phenomena, or the transfer of energy and heat in atmosphere. These studies may address the interaction of ozone and other atmospheric constituents, or uncover new information on global warming and its possible effects. Research scientists may work for organizations such as the National Center for Atmospheric Research (NCAR) in Boulder, Colorado, a facility dedicated to providing state-of-the-art research tools to the atmospheric sciences community. NCAR is home to nine groups and divisions, addressing different facets of the composition and behavior of the atmosphere. These include divisions focusing on atmospheric chemistry, atmospheric technology, mesoscale and microscale meteorology, climate and global dynamics, and scientific computing; a high-altitude observatory; an environmental and societal impacts group; and a research applications program. NCAR also offers an advanced study program, which sponsors between ten and fifteen new postdoctoral scientists each year.

Research meteorologists, also called atmospheric scientists, have usually completed an advanced degree in meteorology, atmospheric sciences, physics, chemistry, or another physical science and tend to be excellent theoreticians or practical scientists. They are usually employed by government laboratories or universities and are less likely to do shift work, except in the case of field experiments. Research meteorologists usually specialize in a particular meteorological discipline such as atmospheric chemistry, atmospheric physics, satellite meteorology, dynamic meteorology, or the improvement of forecast models. Many of these areas overlap, however, requiring researchers to be well versed in all areas of atmospheric science and to collaborate often with people outside their specialty.

An estimated 25 percent of all meteorologists are involved in atmospheric research in the United States. Many of these people work in the National Oceanic and Atmospheric Administration (NOAA) Environmental Research Laboratories, while others conduct research for the National Aeronautics and Space Administration (NASA). Other atmospheric researchers may be affiliated with universities or cooperative institutes across the country.

Other Careers

Private sector areas such as consulting are experiencing high growth and offer potential opportunities for persons with a meteorological background. Meteorologists who work as consultants may address such issues as the effect of a new factory on regional air quality or the best place to construct a new airport, or they may investigate past weather events for insurance companies or public utilities. Air quality is perhaps the most privatized component of meteorology. In this age of permitting, industries often employ meteorologists or hire consultants to assist them in these processes. Many urban areas have pollution control districts or other agencies, boards, or councils responsible for making decisions for the region based on air quality monitoring and modeling.

Several private sector meteorologists work in an operational capacity, providing specially tailored weather products and weather information to a variety of commercial clients. AccuWeather, employing about ninety-three full-time meteorologists, Kavouras DTN, with fifty meteorologists on staff, and WSI, with sixty-five degreed meteorologists in the United States and United Kingdom, are among the largest of these weather information companies. For persons providing consulting services, the American Meteorological Society offers a Certified Consulting Meteorologist accreditation based on formal education requirements, multiple years of experience in the field, and an examination demonstrating meteorological knowledge. As of August 2000, just over 600 meteorologists had received this certification.

Training in meteorology may also provide a suitable background for careers in education. A bachelor's degree in meteorology combined with an education curriculum is excellent preparation for teaching high school math or science, while a person with a master's degree may be able to teach at the community college level. A professorship at a college or university typically requires a Ph.D. and, in many cases, substantial experience in research and related areas.

EDUCATION AND TRAINING

Studying Meteorology

Meteorology is the study of the movement and interaction of molecules in the vast medium of the atmosphere. These movements and interactions are described by the laws of physics, which are written in the language of mathematics. A high level of mathematical understanding, including courses in calculus and differential equations, is important in preparing for a meteorological career. Students interested in such a path can begin preparations in high school by taking every math course available. Many of the mathematical computations necessary to describe the state of the atmosphere can be completed only with computers, so knowledge of computer science and basic programming techniques is also very useful.

Communication skills, including both writing and speaking, are as essential to meteorology as to any other career. Whether in broadcasting, research, teaching, or industry, a meteorologist must be able to communicate clearly and effectively to both fellow professionals and individuals outside the field. For those considering a career in research, a foreign language such as German, Russian, or French can be useful for keeping up to date on progress and breakthroughs worldwide. Courses in government affairs and policy are a good supplement for students interested in air quality work or climate issues.

A list of meteorology and atmospheric science programs in the United States is provided at the end of this chapter, and the American Meteorological Society publication *Curricula in the Atmospheric, Oceanic, Hydrologic, and Related Sciences* contains detailed information on undergraduate and graduate programs. Educational programs in meteorology vary greatly, and the path chosen plays a significant role in determining the types of jobs available upon completion of the degree. Some programs are very broad based, offering courses from forecasting to atmospheric chemistry to climate change, while others tend to focus on a particular component, such as agricultural meteorology or air quality and pollution control. The movement toward increased private sector employment had led some universities to offer more specialized options for meteorology majors. At a 1998 meeting of the heads of academic departments of meteorology and related sciences, examples of some of these alternatives were discussed. Among them are Pennsylvania State University's options in computer science, air quality, climatology, hydrometeorology, weather communications, and teaching, and the University of Oklahoma's concentrations in business, architecture, or computer science. Millersville University of Pennsylvania also offers multiple career tracks, with options in chemistry, broadcast communications, computer science, and natural hazards and emergency management.

A bachelor's degree in meteorology can be the most direct way to a forecasting career, especially for persons interested in working for the weather service or in broadcast meteorology. Twenty-four semester hours of meteorology course work, including 6 hours of weather analysis and prediction and 2 hours of remote sensing and instrumentation, are the minimum requirement for employment with the federal government. Course work must also include calculus, differential equations, 6 hours of physics, and at least 9 hours of courses from supporting subjects, such as chemistry, computer science, statistics, physical oceanography, or climatology. Graduate school can open the door to many more professional opportunities and is virtually mandatory for a career in research. Whether at the undergraduate or graduate level, course work in biology, ecology, oceanography, hydrology, or geophysics can be a good complement to a meteorology degree, especially for persons interested in climate or other multidisciplinary applications.

Internships

Whether a student's interest in meteorology is focused on forecasting, research, or other career areas, internships can be rewarding educational experiences. These opportunities are a wonderful way to explore meteorological careers and work environments and provide valuable work experience to give job-seeking graduates an edge in their search. Many colleges and universities have programs in place allowing students to receive credit for work performed in industry or outside laboratories and can assist students in finding internship opportunities in broadcasting, research, meteorological consulting, or other areas. The National Weather Service offers cooperative opportunities for meteorology students to gain work experience and a beginning on the federal government's career track. Several research institutions offer summer programs for students interested in pursuing these career tracks. A small sampling of these programs includes opportunities at NASA's Goddard Space Flight Center in Maryland, Argonne National Laboratories in Illinois, Oak Ridge Research Laboratories in Tennessee, Los Alamos National Laboratory in New Mexico, and the National Center for Atmospheric Research. University career offices or department faculty can be good sources of information concerning many of these programs. In addition, most meteorological employment listings, especially those provided by the American Meteorological Society, provide thorough information about summer opportunities in various sectors of meteorology.

Martin Venticinique, who completed a meteorology degree at Metropolitan State College of Denver, credits his internship experiences with helping him land his current position as an on-air meteorologist at Channel 13 KRDO-TV in Colorado Springs, Colorado. While pursuing his degree, Venticinique spent two and a half years involved in research with NCAR's Mesoscale and Microscale Meteorology Division. This experience led to an offer of full-time employment with the same group after graduation. He also spent a year as an intern for Denver ABC affiliate KMGH-TV. "An internship is absolutely essential for persons interested in a broadcast career," he says. Television stations hiring broadcast meteorologists require applicants to submit a tape of themselves on camera, which is usually circulated by a talent house. According to Venticinique, "An internship gives you the opportunity to introduce yourself to the technical people, station director, and others. Their help and critique can be invaluable in the production of your tape." In small markets, an internship may also lead directly to a broadcasting job.

Internships are far and away the best method of experiencing of the true nature of meteorological employment. As seasoned intern Venticinique puts it, "You never know what something is like until you are a part of it." Meteorological employment is forecast to become more competitive, meaning that only the very best people will be selected for the best positions. Internships, excellent vehicles through which to gain work experience, also become essential media by which a person's capabilities can be demonstrated.

Pursuing a Career in Meteorology

Several resources are available for persons interested in meteorology careers. Two of the best are the American Meteorological Society and the National Weather Association. Colleges and universities offering meteorological degrees often have student chapters with low dues. These chapters sponsor speakers from the industry and provide a good network for students who are making the transition into the workforce.

Finding and talking to people already in the field can be an excellent way of gleaning information. This section examines the paths that three individuals chose to follow in pursuing their meteorological careers. The biographies of many people who chose research careers are included in Chapter 5 and can be used to provide information about the research career path, which usually requires graduate education and postdoctoral experience. The persons profiled below work for the National Weather Service, a television station, and in the private sector. These individuals have many things in common: all have explored different types of opportunities within meteorology; all have been strongly devoted to seeking out internships and additional educational activities; and, perhaps largely as a result of this, all have been successful in attaining the positions they were seeking.

Coleen Decker first became interested in meteorology after creating a model of the hydrologic cycle when she was in sixth grade. Her mother, a scientist, encouraged her to turn her interest into a career, and today Decker works at the National Weather Service Forecast

Office in Hastings, Nebraska. On her way to becoming a meteorologist, she took advantage of many opportunities to learn about the different types of careers available in this field. She started by joining the Air National Guard Weather program, which provided part-time experience as well as money to go to college. While in college, she interned with a television station and participated in the National Weather Service's Student Cooperative Program. As a co-op student for three years, she learned almost all of the things she needs in her current job and speaks very highly of the experience.

After graduation, Decker worked as a weather observer for the Federal Aviation Administration before joining the National Weather Service. Her duties in her current position include issuing forecasts and warnings for central Nebraska and north-central Kansas, as well as making sure that all meteorological data that come into the office are representative and accurate. She also maintains climatological records dating back over 100 years for the Hastings office and two others.

Severe weather events, including tornadoes, windstorms, and blizzards, are common in Nebraska and Kansas, and during these times Decker and her coworkers are busy issuing short-term forecasts for time periods of two hours or less and coordinating with in-the-field weather observers and emergency managers. Because Decker and other National Weather Service meteorologists work shifts that can change as often as three times per week, everyone in the office gets the opportunity to deal with these extreme events. But shift work does have its trade-offs, including having to work holidays and weekends and problems in scheduling time for family or other events.

Decker's advice to aspiring meteorologists is this: "Study math, math, and more math. Then, be sure that you can take the numbers and explain what they represent in plain English. That's what it's all about. Anyone can rip out a long string of statistics, but the meaning and quality of what we say is what saves lives." Decker also points out that training does not stop with a college degree. With three years of after-college experience behind her, she estimated that it would take two more years of in-house National Weather Service training before she would even be eligible to apply for a forecaster designation.

At the time of this writing, Decker was doing National Weather Service training to attain forecaster status and was also working to be commissioned as an Air Force weather officer. She says that the ever-changing nature of meteorology keeps her interested and motivated, and she expresses hope of eventually sinking her teeth into some research and possibly working toward a science and operations officer position. She also shared the following story about why she chose the path she did and why she has stuck with it:

> Much to my surprise, I had a visitor come in to talk to me yesterday morning after I'd finished my first midnight shift. I wasn't expecting anyone, so it really threw me off guard when a bright-eyed nine-year-old stood waiting in our conference room at a quarter to eight. His pride was seeping through his pores and he stood eagerly clutching a treasure. Going back a few weeks, this same young man called on the phone to request some pamphlets or brochures about weather. He was preparing a report for his 3rd grade science class and needed some reference material. I had prepared a package and enclosed a personal letter with my name on it. To my delight, he had stopped by just to share his project with me and show off the blue ribbon he earned at the science fair. He professionally explained how he compiled all the data into bar graphs and pie charts using a spreadsheet. Then he showed me some of his favorite tornado graphics. Finally, we shared stories about the "coolest" storms we've ever seen. I gave him a lapel pin to wear as a testimony to his efforts and his interests.
>
> After his parents escorted him to their minivan, I thought about the look on his face and the genuine love he had for weather. And you know, it was just like looking in the mirror. He encompasses the next generation of meteorologists. For a long time to come, that will serve as a reminder of why I work all these ghastly midnight shifts, why I spent so much time studying calculus, and why I still find beauty in the mamatus clouds of a thunderstorm. That is what inspired me to become a meteorologist.

Pam Daale, a meteorologist with KMGH-TV in Denver, Colorado, started on her career path in a slightly different way. Her teachers suggested accounting, but after two years of what she thought was the most boring subject in the world, she decided to pursue her other interests. "I just wanted to go into a field I'd enjoy," says Daale. "Meteorology was an obvious choice because I've always been fascinated by weather, especially thunderstorms."

Today Daale forecasts thunderstorms and other weather along Colorado's Front Range and is recognized as a reliable and knowledgeable television meteorologist. As a result of a horseback riding accident when she was sixteen, Pam delivers her forecasts from a wheelchair, but she has never let being a paraplegic interfere with her life or her work. Daale took undergraduate courses in meteorology at Texas A&M University and graduated with a bachelor's degree in meteorology from Iowa State University. While in school, she interned at television and radio stations to gain exposure to broadcast meteorology. Her on-the-air experience began when, even before she had taken a single meteorology class, she found a job reading National Weather Service forecasts for a local radio station in northwest Iowa. While at Texas A&M, Daale did on-the-air forecasts for the PBS station on campus and also worked in radio. When she transferred to Iowa State University, she landed an internship at WOI-TV, the ABC affiliate in Ames, which eventually led to a full-time position. Daale has a Television Seal of Approval from the American Meteorological Society and has worked in Denver, Colorado, since 1993.

Daale brings lots of enthusiasm to her job and loves the challenge of forecasting. The station where she works receives much of the same data and model results used by National Weather Service forecasters, and Daale and her colleagues are given free rein to interpret these numbers. "I love weather. TV is the vehicle for me to play with what I love. It's fun to analyze data and see how things are working, and then to talk about it in a general way," says Daale of her job. She also spends time off the TV screen talking with children and especially loves talking to girls about weather and science. Daale points out that "it's a subject that's easy to get people excited about. And in any conversation about weather, the meteorologist is always the expert." She balances her forecasting responsibilities with many community service projects. At the time of this writing, she was also busy with a four-year-old daughter and a fourteen-month-old son.

John Toohey Morales, Certified Consulting Meteorologist, has worn many hats in his meteorological career. He grew up in Puerto Rico and returned to his birthplace in New York to study atmospheric science at Cornell University. Similar to many other meteorologists, he says this interest started in childhood: "Anything having to do with air and space . . . that's what I was interested in. I remember tracking hurricanes as a pre-teen with much more attention to detail and statistics than other people seemed to care about." During high school, Morales visited the local National Weather Service to find out more about meteorology as career. He spent his college summers interning at the National Weather Service office in San Juan, Puerto Rico. After completing a bachelor's degree in 1984, John began a career with the National Weather Service, working in San Juan; Lake

Charles, Louisiana; and Washington, D.C. He worked his way to forecaster in charge and then accepted a position as chief of the South American desk at the National Centers for Environmental Prediction. In 1988, while at the National Hurricane Center in Florida, Morales completed postgraduate courses in tropical meteorology at the University of Miami.

Morales left government service in 1991 and founded ClimaData Corporation, which provides weather information and forecasts, often in the Spanish language, to clients in the United States and in the Caribbean. The information includes daily faxed forecasts and severe weather outlooks for Intel, Johnson & Johnson, and other clients in Puerto Rico. ClimaData provides game-day forecasts for the Florida Marlins, as well as weather information for hotels and vacationers in the Caribbean islands. Morales is also an associate of Climatological Consulting Corporation, specializing in expert witness testimony for cases in which weather is a factor. He has completed a Certified Consulting Meteorologist accreditation from the American Meteorological Society.

One would think that such successes in government and private sector careers would be enough, but this is still only part of John Toohey Morales's contribution to meteorology. Morales is currently the only meteorologist on Spanish-language television in the United States. His broadcasting career began in 1991 with Noticas 23 at WLTV, the Univisión affiliate in Miami. By 1992, he had become a Univisión network meteorologist, appearing daily across the country and in Latin America. John has won several awards in the Miami market, and also won an Emmy for *48 Horas Antes de la Tormenta*, an information-oriented program detailing steps to take to protect life and property during a hurricane emergency. He holds the National Weather Association's Seal of Approval for Television Weathercasting and is heard on over sixty radio stations in the United States, Puerto Rico, and the Virgin Islands.

EMPLOYMENT OUTLOOK

American Meteorological Society statistics show that about 1,000 new persons enter the meteorological workforce each year. Traditionally, the number of available positions has been virtually equal to the number of new meteorologists. While diligent job seekers are successful in their search, few may be in a position to choose from several offers, and most must remain flexible concerning the desired type of position and part of the country. An undergraduate or master's degree in meteorology or a related science is very adaptable to a position designing new graphics displays for use in television broadcasts, research, and other applications. Jobs in research exist at universities, government labs, and other institutes, mainly for persons holding a master's degree or higher. Many states operate climatology offices staffed by meteorologists and hydrologists for the purpose of maintaining and disseminating regional climate information. A doctoral degree with its accompanying years of training is the typical preparation for any high-level research or university teaching position.

The largest employer of meteorologists has long been the U.S. government, including the National Oceanic and Atmospheric Administration (NOAA), which operates the National Weather Service (NWS). In all, one in four meteorologists is employed by the federal government, and 90 percent of these work in NWS offices around the nation. The Department of Defense and armed forces such as the U.S. Air Force and Navy, the Department of Energy, Department of Agriculture, and the National Aeronautics and Space Administration (NASA) also contribute to the number of government-employed meteorologists.

The 2000–2001 *Occupational Outlook Handbook* offers the most recent information concerning meteorological employment. Compiled by the Bureau of Labor Statistics, this reference predicts slower-than-average employment growth in the field of meteorology through 2008, due largely to cutbacks in the number of meteorologists employed by federal agencies. NWS director John J. Kelly reports that recent restructuring of the NWS has resulted in a decrease from about 230 new professional staff in 1993 to "attrition-based" hiring of fewer than 40 per year through 2001. It is anticipated, however, that the increase in private sector employment may compensate for some of these decreases. The need for private weather services by investors, utilities, construction companies, and farmers is expected to spawn growth in this area, though such growth will be highly dependent on the overall health of the economy.

Salary information associated with meteorological positions is available from a number of sources, including the American Meteorological Society, the Occupational Outlook Handbook, and the Commission on Professionals in Science and Technology (CPST). In 1998, the median salary for persons with a meteorology or atmospheric science degree was $54,430 per year. Beginning salaries for federal meteorologists with a bachelor's degree ranged from $20,600

Figure 6.2. National Weather Service (NWS) meteorologists launch a weather balloon to carry a radiosonde instrument package upward through the atmosphere. The NWS accounts for about 90 percent of the meteorology positions available in the federal government. Courtesy of the National Oceanic and Atmospheric Administration (NOAA).

to $25,500. Master's degree recipients started at $25,500 to $31,200, and those with a Ph.D. could start at $37,100 to $45,200. Advancement in all of these positions can be quite rapid, and the *Occupational Outlook Handbook* reports an average salary of $62,000 for federal meteorologists in 1999. Nine-month teaching appointments typically start at about $33,000, and beginning industry salaries averaged $58,600. Actual earnings vary according to the type of position, with forecasting jobs paying the least, and can depend on the part of the country and cost-of-living estimates. For those interested in broadcast meteorology, the American Meteorological Society reports a salary range from $20,000 and $100,000 per year, with the 1992 average being $46,000.

EMPLOYMENT RESOURCES

A glance through meteorological employment announcements can be an excellent way of finding out more about career options and required skills. A comprehensive listing of meteorological employment resources is provided by Dave Blanchard at <http://mrd3.mmm.ucar.edu/~dob/www/jobs.html>. Other sources for employment information are listed below:

> American Geophysical Union Employment Listings. Published weekly in *Earth Observing System*, and available to subscribers at <http://www.agu.org/cgi-bin/joblistings.cgi>.
>
> American Meteorological Society Employment Announcements. Updated monthly. Available to subscribers only. <http://www.ametsoc.org/AMS/emplyment/>.
>
> EarthWorks. On-line listing of positions in the environmental sciences. Users can also post resumés. <http://ourworld.compuserve.com/homepages/eworks/>.
>
> Employment Resources in the Earth, Atmospheric, and Oceanic Sciences. <http://web.syr.edu/~elwallac/Ewjobs.htm>.
>
> Meteorological Employment Journal. Available by subscription, updated monthly. For more information, see <http://www.swift-site.com/mejjobs/> or write to MEJ, 1102 Plum Tree Dr., Crystal Lake, IL 60014.
>
> MET-JOBS. Electronic mailing of position openings. A free service available at <http://www.eskimo.com/~tcsmith/mail/mj-arc.html>.

National Weather Association. <http://www.nwas.org/jobs.html>.

Sciencewise.com: Opportunities for funding, scholarships, and career development.

Scijobs.org. <http://www.scijobs.org>.

References and Sources of Information

A variety of print and nonprint resources provide additional information about meteorological careers. Some very useful places to begin gathering information are listed below, along with the sources used to compile this chapter.

American Meteorological Society. *Challenges of Our Changing Atmosphere: Careers in Atmospheric Research and Applied Meteorology.* 1993. Available in print on request from AMS or in electronic form at <http://www.ametsoc.org/AMS/pubs/careers.html>.

American Meteorological Society and University Corporation for Atmospheric Research. *Curricula in the Atmospheric, Oceanic, Hydrologic, and Related Sciences.* 2000.

Be a Meteorologist. The Weather Channel. Accessible in electronic form at <http://www.weather.com/learn_more/meteorologist.html#educational>.

Bureau of Labor Statistics. *Occupational Outlook Handbook.* 2000–2001. Published annually by the U.S. Government Printing Office. Available in print or can be accessed electronically at <http://www.bls.gov/ocohome.htm>.

Commission on Professionals in Science and Technology. *Salary and Employment Survey, 1998.* Available in electronic form via Science's NextWave at <http://nextwave.sciencemag.org/survey/>.

Daale, P. Personal communication, 2000.

Decker, C. Personal communication, 2000.

Doswell, Charles A. *What Are You Really Here For Anyway?* A primer on how to be an undergraduate student in physical science or engineering. Available online at <http://www.wildstar.net/~doswell/Outdoor_Images/Student_Book_1.html>.

Douglas, Paul. *Prairie Skies. The Minnesota Weather Book.* Stillwater, MN: Voyageur Press, 1990.

Fiske, Peter S. *To Boldly Go: Practical Career Advice for Geoscientists.* Washington, DC.: American Geophysical Union, 1998.

Kelsey, Jane. *Science.* Lincolnwood, IL: VGM Career Books, 1997. An introduction to careers in various scientific fields, including portraits of people working in meteorology.

Peters, Robert L. *Getting What You Came For: The Smart Student's Guide to Earning a Master's or Ph.D.* New York: Noon Day Press, 1992.

Science Magazine's NextWave. Information on careers and career issues in all fields of science. <http://www.nextwave.org>.

Toohey Morales, J. Personal communication, 2000.

Venticinique, M. Personal communication, 1999.

METEOROLOGY AND ATMOSPHERIC SCIENCES DEGREE PROGRAMS

U.S. and Canadian colleges and universities offering degree programs in the atmospheric, oceanic, hydrologic, and related sciences are listed below, based on a compilation by the American Meteorological Society. The list contains the name and address of the university or college, contact information for the appropriate department, the degrees offered, and the Internet home page, if relevant. Further details on all of the programs can be found in the *2000 Curricula in the Atmospheric, Oceanic, Hydrologic and Related Sciences*, published by the American Meteorological Society.

Air Force Academy, Department of the Air Force
USAF Academy, CO 80840–6238
Department of Economics and Geography, (719) 333–2945
Bachelor of Science
http://www.usafa.af.mil/dfeg/dfeghp.htm

Air Force Institute of Technology
Wright-Patterson AFB, OH 45433
Department of Engineering Physics, (937) 255–2012
Master of Science
http://www.afit.af.mil/schools/en/enp/enphome.html

University of Alabama in Huntsville
Huntsville, AL 35899
Atmospheric Science Department, (205) 922–5754
Master of Science, Doctor of Philosophy
http://www.atmos.uah.edu

University of Alaska
Fairbanks, AK 99775
Department of Physics, (907) 474–7339
Department of Chemistry, (907) 474–7525
Bachelor of Science, Master of Science, Doctor of Philosophy

University of Arizona
Tucson, AZ 85721–0081
Department of Atmospheric Sciences and Institute of Atmospheric Physics, (520) 621–6831
Bachelor of Science, Master of Science, Doctor of Philosophy

University of British Columbia
Vancouver, BC, Canada V6T 1Z2
Atmospheric Science Programme, Department of Geography, (604) 822-5870
Bachelor of Science, Diploma in Meteorology, Master of Science, Doctor of Philosophy
http://www.geog.ubc.ca/atsci

University of California at Davis
Davis, CA 95616
Department of Land, Air and Water Resources, (530) 752-1406
Bachelor of Science, Master of Science, and Doctor of Philosophy in Atmospheric Science
http://www.atm-ucdavis.edu/

University of California at Irvine
Irvine, CA 92697-3100
Department of Earth System Science, (949) 824-8794
Doctor of Philosophy
http://www.ess.uci.edu

University of California, Los Angeles
Los Angeles, CA 90095-1565
Department of Atmospheric Sciences, (310) 825-1217
Bachelor of Science, Master of Science, Doctor of Philosophy

Central Michigan University
Mount Pleasant, MI 48859
Department of Geography, (517) 774-3323
Bachelor of Science

University of Charleston
Charleston, SC 29424
Department of Physics and Astronomy, (803) 953-5593
Bachelor of Arts and Bachelor of Science in Physics and Astronomy with a concentration in Meteorology
http://www.cofc.edu/~physics/physdept.html
Environmental Studies Masters Program, (803) 953-8288
Master of Science
http://www.cofc.edu/~lindnerb/mesphysicstrack.html

University of Chicago
Chicago, IL 60637
Department of the Geophysical Sciences, (773) 702-8101
Bachelor of Arts, Bachelor of Science, Master of Sciences, Doctor of Philosophy
http://geosci.uchicago.edu

University of Colorado at Boulder
Boulder, CO 80309-0311
Program in Atmospheric and Oceanic Sciences, (303) 492-6633
Undergraduate minor, Master of Science, Doctor of Philosophy
http://paos.colorado.edu/

Colorado State University
Fort Collins, CO 80523
Department of Atmospheric Science, (970) 491-8360
Master of Science, Doctor of Philosophy
http://www.atmos.colostate.edu

Cornell University
Ithaca, NY 14853
Atmospheric Science Program, (607) 255-3034
Bachelor of Science, Master of Science, Doctor of Philosophy
http://met-www.cit.cornell.edu/met_home.html

Creighton University
Omaha, NE 68178
Department of Atmospheric Sciences, (402) 280-1731
Bachelor of Science, Master of Science
Environmental Science Program, (402) 280-3190
Bachelor of Science
http://flare.creighton.edu

Dalhousie University
Halifax, Nova Scotia Canada B3H 4J1
Atmospheric Science Program, (902) 494-2337
Diploma in Meteorology, Master of Science, Doctor of Philosophy
Department of Oceanography, (902) 494-3557
Master of Science, Doctor of Philosophy
http://www.phys.ocean.dal.ca

University of Delaware
College of Marine Studies, Lewes, DE 19958

Oceanography Program, (302) 645-4212
Master of Science in Marine Studies, Doctor of Philosophy in Oceanography

University of Denver
Denver, CO 80208
Department of Biological Sciences, (303) 871-3661
Department of Chemistry and Biochemistry, (303) 871-2436
Department of Engineering, (303) 871-2102
Department of Geography, (303) 871-2513
Department of Physics and Astronomy, (303) 871-2238
Bachelor of Arts, Bachelor of Science, Master of Science, Doctor of Philosophy

Drexel University
Philadelphia, PA 19104
School of Environmental Science, Engineering and Policy, (215) 895-2265
Master of Science, Doctor of Philosophy

Florida Institute of Technology
Melbourne, FL 32901
Division of Marine and Environmental Systems, (407) 674-8000 ext. 8096
Bachelor of Science, Master of Science, Doctor of Philosophy

Florida State University
Tallahassee, FL 32306-4520
Department of Meteorology, (850) 644-6205, http://met.fsu.edu
Department of Oceanography, (850) 644-6700, http://ocean.fsu.edu/
Geophysical Fluid Dynamics, Ph.D. Program, (850) 644-5594
Bachelor of Science, Master of Science, Doctor of Philosophy

Georgia Institute of Technology
Atlanta, GA 30332-0340
School of Earth and Atmospheric Sciences, (404) 894-3955
Bachelor of Science, Master of Science, Doctor of Philosophy

University of Guelph
Guelph, Ontario Canada N1G 2W1
Department of Land Resource Science, (519) 824-4120
Master of Science, Doctor of Philosophy
http://www.uoguelph.ca/lrs

Harvard University School of Public Health
Boston, MA 02115
Environmental Science and Engineering Program, Department of Environmental Health, (617) 432–1170
Master of Science, Doctor of Science

University of Hawaii
Honolulu, HI 96822
Department of Meteorology, (808) 956–8775
Bachelor of Science, Master of Science, Doctor of Philosophy
http://lumahai.soest.hawaii.edu/Dept/meteorology/index.html
Department of Oceanography, (808) 956–7633
Master of Science, Doctor of Philosophy
http://www.soest.hawaii.edu/

University of Ilinois at Urbana-Champaign
Urbana, IL 61801
Department of Atmospheric Sciences, (217) 333–2046
Master of Science, Doctor of Philosophy
http://uiatma.atmos.uiuc.edu/
Department of Electrical and Computer Engineering, (217) 333–9705
Bachelor of Science, Master of Science, Doctor of Philosophy
http://www.ece.uiuc.edu/

Indiana University
Bloomington, IN 47405–6101
Department of Geography, Climate and Meteorology Program, (812) 855–6303
Master of Arts, Master of Arts Teaching, Master of Science, Doctor of Philosophy
http://www.indiana.edu/~climate/geoghome.html

Iowa State University
Ames, IA 50011
Department of Geological and Atmospheric Sciences, (515) 294–4758
Bachelor of Science, Master of Science, Doctor of Philosophy

Jackson State University
Jackson, MS 39217–0460
Department of Physics and Atmospheric Sciences, (601) 968–7012
Bachelor of Science in Meteorology, Physics, and Doctorate in Environmental Science
http://santa.jsums.edu/

Johns Hopkins University
Baltimore, MD 21218-2687
Department of Earth and Planetary Sciences, (410) 516-7034
Bachelor of Arts, Master of Arts, Doctor of Philosophy
http://www.jhu.edueps.edu

University of Kansas
Lawrence, KS 66045
Department of Physics and Astronomy, (785) 864-4626
Bachelor of Science, Master of Science
http://www.phsx.ukans.edu

Kean University
Union, NJ 07083
Department of Geology and Meteorology, (908) 527-2064
Bachelor of Arts, Master of Arts
http://www.hurri.kean.edu

Louisiana State University
Baton Rouge, LA 70803
Department of Oceanography and Coastal Sciences, (504) 388-6308
Master of Science, Doctor of Philosophy
www.lib.lsu.edu/oceanography/index.html

Lyndon State College
Lyndonville, VT 05851
Department of Meteorology, (802) 626-6254
Bachelor of Science in Meteorology, minor in Meteorology
http://apollo.lsc.vsc.edu

University of Maryland
College Park, MD 20742-2425
Department of Meteorology, (301) 405-5391
Master of Science, Doctor of Philosophy
http://www.meto.umd.edu

Massachusetts Institute of Technology
Cambridge, MA 02139
Program in Atmospheres, Oceans, and Climate, (617) 253-4464
Master of Science, Doctor of Philosophy, Doctor of Science
http://www-paoc.mit.edu/

University of Massachusetts at Lowell
Lowell, MA 01854
Department of Environmental, Earth and Atmospheric Sciences, (978) 934–3900
Bachelor of Science, Master of Science
http://storm.uml.edu

McGill University
Montreal Canada H3A 2K6
Department of Atmospheric and Oceanic Sciences, (514) 398–3764
Bachelor of Science, Diploma, Master of Science, Doctor of Philosophy
http://zephyr.meteo.mcgill.ca

Metropolitan State College of Denver
Denver, CO 80217
Department of Earth and Atmospheric Sciences, (303) 556–3143
Bachelor of Science
http://clem.mscd.edu/~eas

University of Miami
Miami, FL 33149–1098
Division of Meteorology and Physical Oceanography, (305) 361–4057
Bachelor of Science, Meteorology and Applied Mathematics
Master of Science and Doctor of Philosophy in Meteorology and Physical Oceanography
http://www.rsmas.miami.edu

University of Michigan
Ann Arbor, MI 48109–2143
Department of Atmospheric, Oceanic and Space Sciences, (734) 764-3335
Bachelor of Science, Master of Science, Master of Engineering, Doctor of Philosophy
http://www.aoss.engin.umich.edu

Millersville University of Pennsylvania
Millersville, PA 17551–0302
Department of Earth Sciences, (717) 872–3289
Bachelor of Science, Bachelor of Arts
http://snowball.millersv.edu

University of Minnesota
St. Paul, MN 55108
Department of Soil, Water, and Climate, (612) 625–1244

Department of Geography, (612) 625–6080
Master of Sciences, Doctor of Philosophy

Mississippi State University
Mississippi State, MS 39762
Department of Geosciences, (601) 325–3915
Bachelor of Science, Master of Science, correspondence courses
http://www.msstate.edu/Dept/GeoSciences

University of Missouri at Columbia
Columbia, MO 65211
Department of Soil and Atmospheric Sciences, (573) 882–6591
Bachelor of Science, Master of Science, Doctor of Philosophy
http//www.phlab.missouri.edu/wxcat

University of Missouri at Rolla
Rolla, MO 65409–0430
Physics Department, (573) 341–4363
Master of Science, Doctor of Philosophy

Naval Postgraduate School
Monterey, CA 93943–5000
Department of Meteorology, (408) 656–2516/2517
Master of Science, Doctor of Philosophy
http://www.met.nps.navy.mil
Department of Oceanography, (408) 656–2673/2552
Master of Science, Doctor of Philosophy
http://www.oc.nps.navy.mil

University of Nebraska at Lincoln
Lincoln, NE 68588–0340
Department of Geosciences, (402) 472–2663
Bachelor of Science Meteorology/Climatology
Bachelor of Science, Master of Science, and Ph.D. in Geosciences
http://zephyr.unl.edu
School of Natural Resource Sciences, (402) 472–4915
Master of Science, Doctor of Philosophy
http://www.ianr.unl.edu/cgi/home.pl

University of Nevada at Reno, Desert Research Institute
Reno, NV 89557
Department of Physics, (702) 784–6792, Atmospheric Sciences Program, (702) 677–3192
Master of Science, Doctor of Philosophy in Atmospheric Science
http://www.dri.edu

New Mexico Institute of Mining and Technology
Socorro, NM 87801
Department of Physics, (505) 835-5328
Master of Science, Doctor of Philosophy
http://www/ee/nmt.edu/physics/index/html

State University of New York at Albany
Albany, NY 12222
Department of Earth and Atmospheric Science, (518) 442-4556
Bachelor of Science, Bachelor of Arts, Master of Science, Doctor of Philosophy
http://www.atmos.albany.edu

State University of New York at Brockport
Brockport, NY 14420-2936
Department of the Earth Sciences, (716) 395-2636
Bachelor of Science, Bachelor of Arts
http://www.weather.brockport.edu

State University of New York Maritime College
Bronx, NY 10465
Science Department, (718) 409-7365
Bachelor of Science in Marine Environmental Science

State University of New York at Oneonta
Oneonta, NY 13820-4015
Department of Earth Sciences, (607) 436-3707; Water Resources Program, (607) 436-3064; Meteorology Program, (607) 436-3069
Bachelor of Arts, Bachelor of Science, Master of Arts in Earth Science

State University of New York at Oswego
Oswego, NY 13126
Department of Earth Sciences, (315) 341-2799 or 341-3065
Bachelor of Arts, Bachelor of Science

State University of New York at Stony Brook
Stony Brook, NY 11794-5000
Institute for Terrestrial and Planetary Atmospheres, (516) 632-8009
Bachelor of Science, Master of Science, Doctor of Philosophy
Marine Sciences Research Center, (516) 632-8701
Master of Science, Doctor of Philosophy
http://www.msrc.sunysb.edu/

University of North Carolina at Asheville
Asheville, NC 28804-8511
Department of Atmospheric Sciences, (704) 251-6149
Bachelor of Science
http://www.atms.unca.edu

North Carolina State University
Raleigh, NC 27695-8208
Department of Marine, Earth and Atmospheric Sciences, (919) 515-3711
Bachelor of Science, Master of Science, Doctor of Philosophy
http://www2.ncsu.edu/ncsu/pams/meas/meas_home.html

University of North Dakota
Grand Forks, ND 58202-9006
Department of Atmospheric Sciences, (701) 777-2184
Bachelor of Science

Northeast Louisiana University
Monroe, LA 71209
Department of Geosciences, (318) 342-1878
Bachelor of Science
http://www.net2.nlu.edu/~geos/geoscience.html

University of Northern Colorado
Greeley, CO 80639
Department of Earth Sciences, (970) 351-2647
Bachelor of Arts, Master of Arts

Northern Illinois University
DeKalb, IL 60115
Department of Geography, (815) 753-0631
Bachelor of Science, Master of Science

Northland College
Ashland, WI 54806-3999
Department of Physics and Earth Sciences, (715) 682-1301
Bachelor of Science
http://www.northland.edu

Nova Southeastern University
Dania, FL 33004
Oceanographic Center, (954) 920-1909
Bachelor of Science, Master of Science, Doctor of Philosophy
http://www.nova.edu/ocean

Ohio State University
Columbus, OH 43210-1361
Atmospheric Sciences Program, (614) 292-2514
Bachelor of Science, Master of Science, Doctor of Philosophy

University of Oklahoma
Norman, OK 73019
School of Meteorology, (405) 325-6561
Bachelor of Science, Master of Science, Doctor of Philosophy

Old Dominion University
Norfolk, VA 23529-0276
Department of Ocean, Earth, and Atmospheric Sciences, (757) 683-4285
Master of Science, Doctor of Philosophy

Oregon State University
Covallis, OR 97331-5503
College of Oceanic and Atmospheric Sciences, (541) 737-3504
Master of Arts, Master of Sciences, Doctor of Philosophy
http://www.oce.orst.edu

Pennsylvania State University
University Park, PA 16802
Department of Meteorology, (814) 865-0478
Bachelor of Science, Master of Science, Doctor of Philosophy
http://www.met.psu.edu

Plymouth State College
Plymouth, NH 03264
Natural Science Department, (603) 535-2325
Bachelor of Science
http://vortex.plymouth.edu/

Princeton University
Princeton, NJ 08544-0710
Department of Geological and Geophysical Sciences, (609) 258-6571
Doctor of Philosophy

Purdue University
West Lafayette, IN 47907
Department of Earth and Atmospheric Sciences, (765) 494-3258
Bachelor of Science, Master of Science, Doctor of Philosophy
http://meteor.atms.purdue.edu

University of Rhode Island
Narragansett, RI 02882
Graduate School of Oceanography, (401) 874-6222
Master of Science, Doctor of Philosophy
http://www.gso.uri.edu

Rutgers University—Institute of Marine and Coastal Sciences
New Brunswick, NJ 08901-8521
Department of Marine and Coastal Sciences, (732) 932-6555
Bachelor of Science, in Marine Sciences, Master's and Doctor of Philosophy in Oceanography
http://marine.rutgers.edu

Rutgers—The State University of New Jersey
New Brunswick, NJ 08901-8551
Department of Environmental Sciences, (732) 932-9185
Bachelor of Science, Master of Science, Doctor of Philosophy
http://envsci.rutgers.edu

Saint Cloud State University
St. Cloud, MN 56301-4498
Department of Earth Sciences, (612) 255-3260
Bachelor of Science, Bachelor of Arts
http://coyote.mcs.stcloud.edu:80/dept/earth.science/meteorology/

Saint Louis University
St. Louis, MO 63103-2010
Department of Earth and Atmospheric Sciences, (314) 977-3115
Bachelor of Arts, Bachelor of Science, Master of Professional Meteorology, Master of Science, Doctor of Philosophy
http://www.eas.slu.edu

San Francisco State University
San Francisco, CA 94132
Department of Geosciences, (415) 338-2061
Bachelor of Arts in Science, Master of Science
http://tornado.sfsu.edu/geosciences/geosciences.html

San Jose State University
San Jose, CA 95192-0104
Department of Meteorology, (408) 924-5200
Bachelor of Science, Master of Science

University of Saskatchewan
Saskatoon, Saskatchewan S7N 5E2 Canada
Institute of Space and Atmospheric Studies, (306) 966-6401
Master of Science, Doctor of Philosophy

Scripps Institution of Oceanography, University of California San Diego
La Jolla, CA 92093-0208
Department of the Scripps Institution of Oceanography, (619) 534-3208
Master of Science, Doctor of Philosophy
http://sio.ucsd.edu

University of South Alabama
Mobile, AL 36688
Department of Geology and Geography, (334) 460-6381
Bachelor of Science
http://www.cwrc.geo.usouthal.edu

South Dakota School of Mines and Technology
Rapid City, SD 57701-3995
Department of Atmospheric Sciences, (605) 394-2291
Bachelor of Science, Master of Science, Doctor of Philosophy
http://www.ias.sdsmt.edu/metro/

Texas A&M University
College Station, TX 77843-3150
Department of Meteorology, (409) 845-7671
Bachelor of Science, Master of Science, Doctor of Philosophy
http://www.met.tamu.edu

Texas Tech University
Lubbock, TX 79409-2101
Department of Geosciences, (806) 742-3113
Master of Science, Doctor of Philosophy
http://www.ttu.edu:80/~geosc/
Department of Civil Engineering, (806) 742-3476
Master of Science, Doctor of Philosophy
http://www.ce.ttu.edu/wind

University of Toronto
Toronto, Ontario, Canada M5S 1A7
Department of Physics, (416) 978-5205
Bachelor of Science, Master of Science, Doctor of Philosophy
http://www.atmosp.physics.utoronto.ca

U.S. Naval Academy
Annapolis, MD 21402
Oceanography Department, (410) 293-6550
Bachelor of Science
http://www.nadn.mil/oceanography/homepage.html

Utah State University
Logan, UT 84322
Department of Plants, Soils, and Biometeorology, (801) 797-2233
Master of Science in Biometeorology, Doctor of Philosophy in Biometeorology
http://psb.usu.edu

University of Utah
Salt Lake City, UT 84112-0110
Department of Meteorology, (801) 581-6136
Bachelor of Science, Master of Science, Doctor of Philosophy
http://www.met.utah.edu

Valparaiso University
Valparaiso, IN 46383
Department of Geography and Meteorology, (219) 464-5140
Bachelor of Science (Meteorology), Bachelor of Arts (Broadcast Meteorology)
http://www.valpo.edu/geomet/vumete.html

University of Virginia
Charlottesville, VA 22903
Department of Environmental Sciences, (804) 924-7761
Bachelor of Arts, Master of Science, Doctor of Philosophy

University of Washington
Seattle, WA 98195
School of Oceanography, (206) 543-5060
Bachelor of Arts, Bachelor of Science, Master of Science, Doctor of Philosophy
Department of Atmospheric Sciences, (206) 543-4250
Bachelor of Science, Master of Science, Doctor of Philosophy
http://www.atmos.washington.edu

Western Connecticut State University
Danbury, CT 06810
Department of Physics and Astronomy, (203) 837–8670
Bachelor of Arts in Meteorology

Western Illinois University
Macomb, IL 61455–1396
Department of Geography, (309) 298–1648
Bachelor of Science
http://www.ECNet.Net/users/migeog/wiu/

William Marsh Rice University
Houston, TX 77005–1892
Department of Space Physics and Astronomy, (713) 527–4939
Bachelor of Arts, Master of Arts, Doctor of Philosophy

University of Wisconsin-Madison
Madison, WI 53706
Department of Atmospheric and Oceanic Sciences, (608) 262–2828
Bachelor of Science, Master of Science, Doctor of Philosophy
http://www.meteor.wisc.edu
Oceanography and Limnology Graduate Program, (608) 263–3264
Master of Science, Doctor of Philosophy

University of Wisconsin-Milwaukee
Milwaukee, WI 53201
Department of Geosciences, (414) 229–4561
Bachelor of Science, Master of Science, Doctor of Philosophy

Woods Hole Oceanographic Institution
Woods Hole, MA 02543
Physical Oceanography Department, Applied Ocean Physics and Engineering, Geology and Geophysics Department, Chemistry Department, Biology Department, (508) 457–2200
Master of Science, Doctor of Philosophy, Doctor of Science
http://www.whoi.edu

University of Wyoming
Laramie, WY 82071
Department of Atmospheric Science, (307) 766–3246
Master of Science, Doctor of Philosophy

York University
Toronto, Ontario M3J 1P3 Canada
Department of Earth and Atmospheric Science, (416) 736-5245
Bachelor of Science (Ordinary and Specialized or Combined Honours),
Master of Science, Doctor of Philosophy
http://www.eas.yorku.ca

Chapter Seven

Statistics and Data

This chapter contains information on different aspects of meteorology, including graduation projections for meteorology students, career choices of graduates, and statistics related to weather disasters. It also contains information on funding for meteorology and atmospheric research, in the past as well as projected for the future.

METEOROLOGICAL EMPLOYMENT

Figure 7.1 shows the projected number of bachelor's, master's, and doctoral degree recipients through 2002. Overall, it shows that meteorology is a relatively small field, with only 500 to 600 students receiving bachelor's degrees in a typical year and even fewer receiving a master's or Ph.D.

Figure 7.2 shows the career choices of meteorology and atmospheric sciences graduates by degree. The numbers reflect the career choices of persons receiving a degree between 1995 and 1997. The private sector is currently the largest employer of bachelor's-level graduates, followed closely by the military. A significant number of bachelor's degree recipients pursue further university education. Many master's degree recipients also choose to pursue further education, while others work in either the private sector or the military or in the civilian government. The government is one of the largest employers of Ph.D. recipients.

Figure 7.3 shows a breakdown of the educational level of American

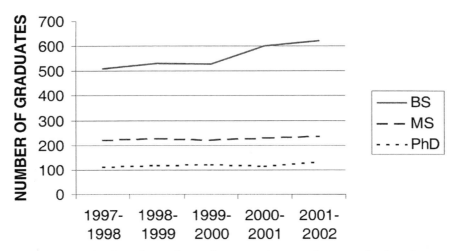

Figure 7.1. Projected number of meteorology and atmospheric sciences degree recipients through 2002, compiled by the American Meteorological Society, based on data from fifty-six schools.

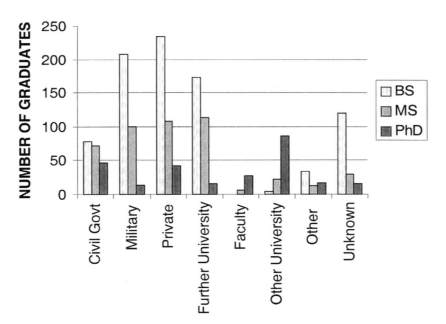

Figure 7.2. Career choices of meteorology/atmospheric sciences graduates, 1995–1997, compiled by the American Meteorological Society, based on data from fifty-two schools.

Statistics and Data 191

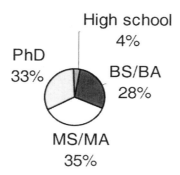

Figure 7.3. Educational levels of American Meteorological Society members, compiled in 1998 as part of the AMS Ten-Year Vision Study, 1998.

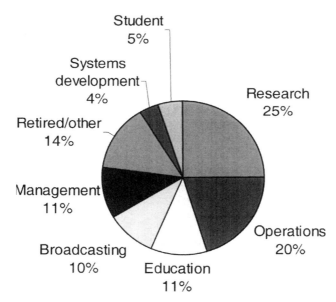

Figure 7.4. Types of employment pursued by American Meteorological Society members. AMS, Ten-Year Vision Study, 1998.

Meteorological Society (AMS) members and indicates an approximately even split among bachelor's, master's, and Ph.D. recipients. Of the members participating in the survey, 87 percent held a degree in meteorology or the atmospheric sciences. Oceanography and hydrology were the most common other subject areas.

Figure 7.4 illustrates the breakdown of types of employment pur-

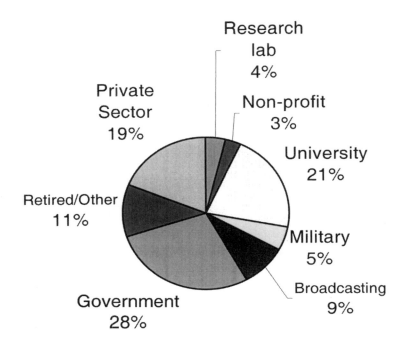

Figure 7.5. Employment of American Meteorological Society members. AMS, Ten-Year Vision Study, 1998.

sued by AMS members. Research is the largest portion of the spectrum, followed closely by operations, or weather forecasting. Other meteorologists work as teachers or broadcasters, or in management. The employers of meteorologists are shown in Figure 7.5. The government is the largest employer of AMS members, employing a large number of research personnel. Universities are the second largest source of employment, followed closely by the private sector. Only one of every ten meteorologists works in broadcasting.

Table 7.1 shows starting salaries in the atmospheric sciences based on information from the 2000–2001 *Occupational Outlook Handbook*. The values seem low, but readers should keep in mind that compensation varies according to position, with entry-level forecasting jobs typically paying least. In most meteorological employment categories, salaries have increased to keep pace with the rising cost of living in recent years, and can also vary by region of the country according to cost of living differences. At all education levels but especially in federal positions, advancement can be rapid.

Statistics and Data

Table 7.1
Starting Salaries in Meteorology and Atmospheric Science

	Federal Government	Private Sector	
B.S./B.A.	$20,600–25,500	$22,000	
M.S./M.A.	$25,500–31,200	$27,000	
		Teaching (9 mo.)	Industry
Ph.D.	$37,700–47,500	$33,000	$58,600

Source: Based on information from the 2000–2001 Occupational Outlook Handbook.

Budgetary outlays for the major U.S. agencies participating in atmospheric research are shown in Figure 7.6. The year 2001 numbers represent requested amounts; actual allocations are finalized in September of each year. Overall, the numbers have been largely stagnant over the past years, in some cases even showing decreases.

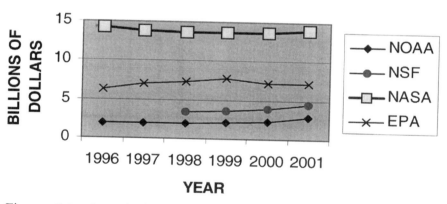

Figure 7.6. Annual budget information for major U.S. agencies participating in atmospheric research. Data compiled by the American Meteorological Society.

WEATHER-RELATED DAMAGE AND FATALITIES

Table 7.2 shows the number of weather-related fatalities in the United States from 1991 to 1995. The totals are strongly influenced by deaths due to extreme temperatures, usually associated with heat waves in major cities. Current meteorological research and forecasting efforts focus on improving our understanding of the atmosphere and its phenomena, thereby reducing these numbers.

Table 7.3 reports weather-related damage in the United States from 1991 to 1995. The figures, given in millions of dollars, clearly show the impacts of Hurricane Andrew in 1992 and the Mississippi River floods in 1993. Many of the dollars budgeted for research and operations, as shown in Figure 7.6, support efforts to help alleviate these damages.

A list of U.S. industries affected or influenced by weather and climate is shown in Table 7.4. Of these industries, finance, insurance, and real estate compose the largest percentage of the gross domestic product (18.4 percent). Retail, transportation, and public utilities are also economically significant industries with a high degree of weather sensitivity.

Table 7.2
Weather-Related Fatalities in the United States, 1991–1995

Event	1991	1992	1993	1994	1995	Total
Extreme temperatures	49	22	38	81	1,043	1,233
Convective storms[a]	144	93	99	155	153	644
Floods	61	62	103	91	90	397
Other[b]	75	40	62	20	51	248
Snow, ice, avalanches	45	64	67	31	17	224
Hurricanes	13	27	2	9	17	68
Marine storms	4	0	0	1	1	6
Total	391	308	371	388	1,362	2,820
Annual average						564

Note: Fatalities represented are only those deemed directly attributable to weather and floods. The number of fatalities in which weather was a contributing factor would be much higher.
[a]Includes tornadoes, thunderstorms, lightning, and hail.
[b]Includes drought, dust storms, rain, fog, strong winds, fire weather, and mud slides.

Source: Office of Meteorology, National Weather Service, *The Atmospheric Sciences: Entering the Twenty-first Century* (Washington, DC: National Research Council, 1998).

Table 7.3
Weather-Related Damage in the United States, 1991–1995 (in $millions)

Event	1991	1992	1993	1994	1995	Total
Hurricanes	1,164	33,661	15	426	5,932	41,148
Floods	874	690	21,288	921	1,250	25,023
Other[b]	1,878	1,932	5,019	893	359	10,081
Convective storms[a]	1,527	1,580	1,086	1,001	2,638	7,832
Snow, ice, avalanches	516	28	602	1,143	111	2,400
Extreme temperatures	224	479	416	52	1,120	2,291
Marine storms	45	31	1	3	2	82
Total	6,228	38,351	28,427	4,439	11,412	88,857
Annual average						17,771

[a] Includes tornadoes, thunderstorms, lightning, and hail.
[b] Includes drought, dust storms, rain, fog, strong winds, fire weather, and mud slides.

Source: Office of Meteorology, National Weather Service, *The Atmospheric Sciences: Entering the Twenty-First Century* (Washington, DC: National Research Council, 1998).

Table 7.4
U.S. Industries Affected or Influenced by Weather and Climate

Industry	Contribution to 1996 Gross Domestic Product ($ billion)	Percent of Gross Domestic Product
Agriculture, fisheries, forestry	115.5	1.9
Construction	222.1	3.7
Transportation, public utilities	529.3	8.8
Retail trade	557.5	9.3
Finance, insurance, real estate	1106.1	18.4
Total	2530.5	42.1

Source: Bureau of Economic Analysis, Department of Commerce, *The Atmospheric Sciences: Entering the Twenty-first Century* (Washington, DC: National Research Council, 1998).

Chapter Eight

Selected Reprints, Documents, and Reports

The previous chapters present information on the major issues under study by meteorologists and atmospheric scientists. Weather and climate affect all aspects of human life, and scientists and policymakers alike must understand the issues well to recommend proper plans for action to protect human lives, property, and well-being. The excerpts reprinted here present scientists' stands on major issues, from weather forecasting to climate change to ozone.

FORECASTING

One of the major issues debated in the weather forecasting forum concerns the privatization of weather services. We include the American Meteorological Society's policy statement on the public and private provision of weather and climate services. The document highlights the need for a public-private partnership in providing weather information and recommends guidelines for how both public and private entities can contribute most effectively to the weather needs of society. This policy statement was adopted by the society's executive committee on July 23, 1999.

The Public/Private Partnership in the Provision of Weather and Climate Services

The American Meteorological Society believes that a strong partnership between the public and private sectors in the provision of weather, climate,

hydrologic, and environmental services is a critical element in the application of our science. A coordinated and cooperative sharing of responsibilities will
- Enhance public safety and the protection of property
- Benefit industry and commerce
- Speed the transfer of new technologies and scientific knowledge to benefit society
- Enhance growth in the private sector and stable support for the public sector
- Best serve the interests of users of weather and climate information

Unprecedented progress in meteorological science and service has been achieved during the twentieth century, in part as a result of extensive public investments in research and technology. Weather forecasting has matured such that many thousands, perhaps countless decisions are made on the basis of weather forecasts each day. Science-based predictions of climate variations up to several seasons in advance can now be made with useful accuracy, as was demonstrated during the 1997–1998 El Niño episode. In parallel, a flourishing private sector has developed, providing tailored weather and climate information to individuals and organizations. The American Meteorological Society anticipates and encourages growth in this sector, as businesses and other organizations increasingly recognize the potential of such information to not only help avoid economic losses from disruptive weather, but also to exploit reliable information on nondisruptive weather to enhance economic performance.

Since the 1930s the private sector has invested extensively in the science, technology, delivery, and marketing of value-added weather products in the United States and abroad. As a result of this investment, the private sector delivers made-to-order weather information that strengthens a business's performance in many ways. The private sector is at the forefront of developing creative products for the media that substantially improve the display and communication of weather information to the general public. Industries subject to environmental regulation by the government depend on the private sector for instruments and real-time input to critical decisions. The transportation industry improves the safety of travelers and their own on-time performance by weaving specialized information from the private sector into the fabric of their operations. These are but a few of today's market-driven innovations from the private sector that have led to the widespread application of weather information by the business community. Tomorrow's innovations will continue to expand the acceptance of the value of weather information into more and more sectors of the global economy.

In addition to the vigorous private sector, the Society also recognizes the importance of the public sector in the provision of weather and climate services for the common good. Benefits from these services to the people of the United States include: the protection of life and property, an ability to cope with environmental problems, and help in making a wide range of weather and climate-based decisions. The public sector also plays an important role in international activities involving the atmosphere, including treaty obligations such as the Safety Of Life At Sea agreement, aviation protocols and procedures under the auspices of the International Civil Aviation Organization, and the World Meteorological Organization of the United Nations.

The observation and monitoring systems, which include constellations of satellites, and networks of radars, buoys, aircraft, and land-based observatories, are by far the most expensive component of the infrastructure. These systems have multiple uses by several public agencies in carrying out their missions associated with weather, climate, water resource management, environmental monitoring, and a host of other governmental responsibilities. Over the last decade, a significant public investment in this infrastructure through the modernization of NOAA's National Weather Service has proven invaluable to both public agencies and the private sector. It has resulted in dramatic gains in observing capabilities, advances in atmospheric understanding, and economically beneficial increases in forecast skill.

The public sector infrastructure-observation system, communication, data processing, and computing systems, and a system for forecasts and warnings—while designed and maintained by the public sector in order to discharge its own responsibilities, is also essential to the private sector in the generation of its products and services. A vigorous and healthy partnership between the private sector and the public sector is therefore of great importance to the expanding use of weather and climate information in the United States. In the spirit of maintaining and enhancing this partnership, the AMS offers the following view of the role for each partner such that the whole is greater than the sum of its parts.

GOVERNMENT

It is the responsibility of NOAA to provide for use by the public of the United States, weather, climate and hydrologic information, and forecasts for the entire country, including the coastal and adjacent ocean areas. The government gives priority to the provision of forecasts and warnings of severe weather, floods, and maritime conditions, and for aviation and wildfire control, which involve the protection of life and property and contribute to the safety and well being of all people.

To carry out its mission, NOAA, in cooperation with other appropriate federal and state agencies:
- Maintains and continually improves a weather, climate and hydrologic observation, communication, data processing, archive, access, and forecasting system that is required to fulfill NOAA's responsibilities
- Cooperates with the private sector to facilitate the further dissemination and interpretation of weather, climate and hydrologic information as a means of increasing the value of such information to society
- Conducts, in cooperation with universities, the private sector, and other institutions, a program of research and development that continually improves the quality of products and services from both the public and private sectors

THE PRIVATE SECTOR

The private sector provides a variety of value-added meteorological products and custom-tailored weather and climate information services, which enhance the basic infrastructure, products and services provided by the public sector. The private sector provides services to the public, and responds to meteorological needs within various economic sectors of society that are sensitive to meteorological, oceanographic, hydrologic, and environmental variations. The private sector leads in the provision of advanced technology for the improvement of the infrastructure.

The private sector performs the following functions within the public/private partnership:
- Disseminates public domain weather and climate information to the public and other users, in cooperation with NOAA and emergency management organizations. In the case of situations where life and property are threatened, the private sector relays public sector warnings and advisories to the public, ensuring that a consistent, unified voice is heard by those affected citizens.
- Produces and delivers valued-added weather, climate, hydrologic, and environmental products and services, and promotes their widest and most productive commercial application, to enhance the efficiency of economic sectors that are sensitive to weather and climate variations.
- Creates technological advances in observations, computing, communications, and any other areas necessary for progress in the science and application of meteorology and related aspects of oceanography and hydrology.

PARTNERSHIP

Frequent and effective communication between partners is vital to a successful partnership. Both public and private sectors should strive for a teamwork approach on such topics as public safety during hazardous weather events, where the public sector has the lead responsibility, and in the enhancing of efficiency in economic sectors sensitive to weather and climate, where the private sector has the lead. Teamwork requires good communication. The public sector should seek out feedback from universities, research institutions, and the private sector regarding their performance, including suggestions for service improvements, and the private sector, universities, and other research institutions should provide that feedback.

SUMMARY

The American Meteorological Society strongly believes that the economic position of U.S. industries impacted by weather and climate will be well served, the private weather/climate sector will experience unprecedented growth, and the general public will continue to benefit as a result of this public/private partnership.

The next selection is an editorial by Roger A. Pielke, Jr., a scientist with the National Center for Atmospheric Research. Pielke's "Six Heretical Notions About Weather Policy" highlights some of the nuances of the public versus private debate. The editorial appeared in the April 2000 issue of *Weatherzine* and provoked much discussion on the topic.

Six Heretical Notions About Weather Policy

When in 1996 freshman Representative Dick Chrysler (R-MI) proposed eliminating the National Weather Service, he declared that the agency was not needed because "I get my weather from the weather channel." Mr. Chrysler's odd view reinforced the perspective of many in the weather community that policy makers have little understanding of meteorological research and operations. On more than one occasion I have heard weather scientists and administrators look with envy to the fortunes of the National Earthquake Hazards Reduction Program that began in 1977 and continues today with about $100 million in annual funding, as well as to the fortunes of programs focused on global change, and more recently, on the carbon cycle. Indeed, the seeming low priority placed on the U.S. Weather Research Program, as measured by appropriated budgets, gives some plausibility to the idea that weather has been less successful in the budget process than other research communities.

The perception that the weather community has been less successful than other communities has led me to take a closer look at its role in the broader environment of science policies and priorities. My initial reaction was that many of the obstacles standing in the way of increased resources for the nation's weather enterprise are to be found inside not outside the community. I rapidly realized that I was quickly entering what some colleagues might call "heretical" territory. So I thought I'd raise these heretical notions, to stimulate debate and discussion, and hopefully to initiate a dialogue on important issues facing the community.

To continue to raise these issues, I present below "Six Heretical Notions about Weather Policy" that I presented twice last year, first to a joint meeting of the National Academy of Sciences Board on Atmospheric Sciences and Climate and the Federal Committee for Meteorological Research and Supporting Services, and then second to the Interdepartmental Committee on Meteorological Services and Supporting Research. Here is what I presented.

Purpose: These "heretical" notions are raised to stimulate debate and discussion. No claim is made as to their relative or absolute truth.

The weather community postulates that improved forecasts will benefit society.

Thus, the logic about what to do is obvious.

To improve forecasts we must advance science.

To advance science we need improved models.

To test and use improved models we need better observations.

To assimilate the better observations and run the improved models we need faster computers.

More funding will enable faster computers, better observations, improved models, and advances in science.

Therefore, more funding for advancements in science, models, observations, and computers are necessary and sufficient to benefit society. A corollary is that the greater the rate of these advancements, the greater the benefits to society. This logic seems so obvious and inescapable that to many, great frustration is sometimes expressed when policy makers in places like Congress and the Office of Management and Budget apparently fail to grasp its self-evidence.

But how might a well-meaning, but scrutinizing, person evaluate this logic? They might raise some "heretical" notions!!

Notion #1. The atmospheric sciences collect more data than is used or can be used in either research or operations. When field programs or satellites are funded, subsequent analysis often is not. This circumstance makes it difficult for people outside the community (and indeed some inside) to understand why more data is needed, and what its ultimate value is in terms of improvements in forecasts as well as opportunities forgone.

Notion #2. Many claim that the forecasts in the United States are the best in the world. At the same time, some folks claim that the Europeans (www.ecmwf.int) have passed us by. Some who say that the United States is keeping pace with the Europeans argue that we have done so because of innovative use of observations (via creative data assimilation techniques and use of scarce computer time). This is tantamount to saying that funding limitations have motivated extra value from existing resources. The bottom line: Do we really know how "good" forecasts have to be, and at what cost?

Notion #3. In any case, public funding for the atmospheric sciences is truly enormous—approximately $2–3 billion is spent on weather and climate research and operations each. When the weather community says that forecasts could improve but only for a small budget increase, one might expect a policy maker to reply: "Great, you should be able to handle that with existing expenditures!"

Notion #4. Much more research is produced than is used, or can be used, in the operational forecast process. Much is "left on the floor." The connections between research and the use of research in operations and ultimately in benefits to decision-makers is poorly understood (www.esig.ucar.edu/socasp/zine/15.html#1). Until the community can link a request for more resources with expected effects on forecasts and ultimately benefits, securing significant additional funding will be difficult.

Notion #5. In any case, improved use and value is in many instances constrained by dated products and a lack of understanding of the needs of users. Remember the case of Grand Forks (www.esig.ucar.edu/redriver/) when a technically accurate forecast was misinterpreted and misused because neither forecasters nor local decision makers understood what it meant. Scientific and technological advances mean little if they are not well incorporated into decision making.

Notion #6. The weather community is so large and full of overlaps and redundancies that no one really knows what the universe looks like. It is difficult for a community like the weather community to speak with one voice, but at a minimum there should be some knowledge of the whole. And there is the destructive public-private debate over roles and responsibilities. Obtaining such knowledge and resolving this debate would greatly enhance credibility when a case is made for more support from the public.

So what needs to be done to better serve the interests of the weather community, the public, and their elected representatives? As a first step these six heresies might be discussed, debated, and perhaps even dismissed by the community. In the process of doing so, the community might find itself better prepared to deal with the views of well-meaning, but scrutinizing public officials.

The American Meteorological Society's policy statement on planned and inadvertent weather modification, which follows, was adopted by the AMS Council on October 2, 1998. It presents information on the status of intentional weather modification efforts, discusses what scientists know about inadvertent changes in local and regional-scale weather due to human activities, and sets out recommendations for further progress in understanding both intentional and accidental modification of the weather.

Planned and Inadvertent Weather Modification

This statement is concerned with the scientific status of planned and inadvertent weather modification on local and regional scales. The Society's policy on global climate change is separate and has previously been presented (*Bull. Amer. Meteor. Soc.*, 72, 57).

1. STATUS OF PLANNED WEATHER MODIFICATION

a. Fog and Stratus Removal

Operations that dissipate supercooled fog and low stratus (clouds containing water droplets at subfreezing temperatures) by seeding with ice-forming agents (e.g., dry ice, liquid nitrogen, compressed air, silver iodide, etc.) have become routine at some airports. The dissipation of warm fogs can be accomplished by more expensive thermal techniques, but this has proven cost effective at only a few major airports.

b. Precipitation Increase

There is statistical evidence that precipitation from supercooled orographic clouds (clouds that develop over mountains) has been seasonally increased by about 10%. The physical cause-and-effect relationships, however, have not been fully documented. Nevertheless, the potential for such increases is supported by field measurements and numerical model simulations.

Some experiments with warm-based convective clouds [bases about 10°C (~50°F) or warmer] involving heavy silver iodide seeding have suggested a positive effect on individual convective cells, but conclusive evidence that such seeding can increase rainfall from multicell storms has yet to be established. Many steps in the physical chain of events are not well understood at this time and have not been documented with observations nor simulated in numerical modeling experiments.

In recent years, the seeding of warm and cold convective clouds with hygroscopic chemical particles to augment rainfall has received renewed attention through model simulations and field experiments. A recent randomized experiment has reported statistical evidence of a rainfall increase that is supported by numerical modeling experiments. Nevertheless, measurements of key steps in the chain of physical events associated with hygroscopic particle seeding are needed to confirm the seeding hypothesis and the range of effectiveness of these techniques in increasing precipitation. Evidence that such seeding can increase rainfall over economically significant areas is not yet available.

There are indications that precipitation changes, either increases or decreases, can also occur at some distance beyond intended target areas (extra-area effects). Improved quantification of the associated hydrological impacts is needed to satisfy public concerns.

c. Hail Suppression

Results of various operational and experimental projects and numerical modeling experiments provide a range of outcomes: some suggest decreases or increases in hail while others have produced inconclusive results. Statistical assessments of certain operational projects indicate successful reduction of crop hail damage, but the physical basis for these results has not yet been established.

d. Severe Storms Modification

No sound physical hypotheses exist for the modification of hurricanes, tornadoes, or damaging winds in general, and no related scientific experimentation has been conducted in the past 20 years. Experiments have been carried out on lightning suppression but have not yielded methods for application.

2. STATUS OF INADVERTENT WEATHER MODIFICATION

There is ample evidence that agricultural and industrial activities modify local and sometimes regional weather conditions. Improved environmental monitoring and atmospheric modeling capabilities have revealed that human activities have significant impacts on meteorological parameters and climatological mechanisms that influence our health, productivity, and societal infrastructure. The environmental impacts of acid rain on structures, vegetation, and lake water quality; of increased anthropogenic pollutants on air quality and visibility; and of urban effects on temperature, humidity, wind, and precipitation have all been well documented. In addition, atmospheric changes that might have passed unnoticed, or been dismissed as inconsequential just a few years ago, are now often found to have broader ramifications; for example, increased cloudiness associated with condensation trails from jet aircraft may modify the radiation budget at the ground.

3. RECOMMENDATIONS

Increasing population, shifting demographics, and the prospect of global climatic change require that food, fiber, and water resources be managed to best alleviate the chronic shortages that are already beginning to manifest themselves. Thus, there is considerable need and benefit to determine the scientific and economic feasibility of cloud-modification methods. Likewise, actions that inadvertently modify weather or climate need to be better understood, quantified, and (if necessary and feasible) mitigated. These are challenging tasks requiring well-focused, long-term efforts. Major breakthroughs in any of these areas are unlikely; progress will more probably continue to be evolutionary.

1) Planned weather modification programs are unlikely to achieve higher scientific credibility until more complete understanding of the physical processes responsible for any modification effect is established and linked by direct observation to the specific seeding methodology employed. Recent improvements in seeding agents, observational facilities and platforms, computer capabilities, numerical models, and physical understanding now permit more detailed examination of clouds and precipitation processes than ever before, and significant advances are consequently possible. Whereas a statistical evaluation is required to establish that a significant change resulted from a given seeding activity, it must be accompanied by a physical evaluation to confirm that the statistically observed change was due to the seeding.

2) Precipitation augmentation through cloud seeding should not be viewed as a drought relief measure. Opportunities to increase precipitation are usually few, if any, during droughts; consequently, the cost of mounting a cloud-seeding operation will far exceed the benefits that may be obtained. A program of precipitation augmentation is more effective in cushioning the impact of drought if it is used as part of a water management strategy on a year-round basis whenever opportunities exist to build soil moisture, to improve cropland, and to increase water in storage.

3) Anthropogenic influences on meteorological and climatological conditions are far reaching and significant. Economic development should be framed with an awareness of how each activity alters environmental processes and atmospheric conditions. Much is known about the physical processes involved in many aspects of inadvertent weather modification, but important questions remain. Continuing research and monitoring related to inadvertent weather modification is required, and the breadth of these studies must be extended to include new knowledge on natural feedbacks and societal

ramifications, which will lead to policy decisions that reduce the chance of severe impacts.

GREENHOUSE GAS EMISSIONS AND CLIMATE CHANGE

Scientists have documented a decades-long increase in greenhouse gases in our atmosphere and have compiled ample evidence that this increase is human caused and will contribute to a warming of our planet's surface. The first selection is an excerpt from a speech that Timothy E. Wirth, U.S. undersecretary for global affairs, delivered to the National Academy of Sciences on November 17, 1997. This speech was released by the Bureau of Oceans and International Environmental and Scientific Affairs, U.S. Department of State.

Remarks at the Forum on International Geosciences at the National Academy of Sciences

In the 1980s, scientific evidence about the pace of global climate change and its consequences led to growing concern among scientists and policy makers. They feared the effects of an unprecedented growth of greenhouse gas emissions in the past century and their projected steep ascent into the next century.

In 1988 the Intergovernmental Panel on Climate Change (the IPCC) was established to provide the scientific understanding necessary to formulate an appropriate policy response to global warming. The IPCC may be the largest international scientific collaboration ever.

Through the IPCC process, more than 2000 of the world's leading climate scientists from more than 150 countries have assessed the available information on climate change and its environmental and economic effects. As you know, they have issued a series of reports, incorporating extensive peer review and a commitment to scientific excellence, that provide the most authoritative and comprehensive information available on the science of climate change.

We believe that the science is compelling:

The chemical composition of the atmosphere is being altered by anthropogenic emissions of greenhouse gases.

The continued buildup of these gases will enhance the natural greenhouse effect and cause the global climate to change.

The balance of evidence suggests that there is a discernible human influence on global climate, in the words of the IPCC. This was the first time that a consensus emerged among leading climate scientists that the world's changing climatic conditions are more than the natural variability of weather.

Nonetheless, uncertainty remains. The scientific community cannot yet tell us precisely how much, when, or at what rate the Earth's climate will respond to greenhouse gas buildup. However, making the best possible estimate based on what is known about the complex climate system, the scientific community believes that current emissions trends—resulting over the next several decades in the effective doubling from preindustrial concentrations of carbon dioxide in the atmosphere—will lead to global temperatures which, on average, are 2 to 6.5 degrees warmer than today, increasing at a rate greater than any known for the past 10,000 years.

Human induced climate change, if allowed to continue unabated, could have profound consequences for the economy and the quality of life of future generations:

Human health will be threatened in a variety of ways. Heat related mortalities may increase, as may deaths resulting from violent weather, such as floods. Diseases that thrive in warmer climates, such as malaria, dengue, yellow fever, and cholera are likely to spread as the range of mosquitoes and other disease-carrying organisms expand. There are projections of between 50 and 80 million additional malaria cases per year as a result of climate change.

A global warming–induced rise in sea level will endanger coastal areas where a large percentage of the world's population lives. A substantial rise in sea level threatens the very existence of some islands, displacing millions of people. Protecting coastal areas from rising sea levels will cost governments billions of dollars.

Even small degrees of climate change will affect agriculture and fish production, leaving many regions of the world vulnerable to food shortages. Climate change is likely to disrupt the water cycle, resulting in more intense droughts in some regions, and floods in other areas. In addition, the quantity and quality of water resources will be stressed, especially in areas that are already arid or semi-arid. Conflict over scarce water supplies may lead to political disruptions.

The IPCC has clearly demonstrated to policy makers that further action must be taken to address this challenge. U.S. policy on climate change flows from this science, and from the understanding that the risks are too great to ignore. We must act now.

Our proposal, as we move ever closer toward the Kyoto Conference in two weeks, has three central components:

We propose that all developed countries have realistic and achievable targets and timetables for significantly reducing their greenhouse gas emissions.

We propose that developing nations advance their commitments undertaken as part of the original Climate Change Treaty and further, agree to a process that will ensure that they will have binding emissions limitation commitments of their own.

And, finally, we propose to establish a system of emissions trading and other market mechanisms that will reduce the costs of limiting emissions, in both developed and developing countries.

Let me address each of these three aspects of our proposal in turn.

I want to start with our ideas for emissions trading and market mechanisms, because we see them as essential to our whole proposal, both environmentally and economically. We believe that these market mechanisms can reduce the costs of implementation significantly, thus enabling us to achieve much greater reductions in emissions in both developed and developing nations.

In the United States, the concept of emissions trading has been successfully used to reduce costs as much as tenfold in meeting the standards set for power plant emissions of sulfur dioxide. In the climate context, we envision that participating nations, and their private-sector companies, would be allowed to trade greenhouse gas emission permits, thus creating the opportunity to reduce emissions where it is cheapest to do so. Such a program could cut the cost of reducing emissions by as much as fifty percent.

It will take some time to design and implement an international emissions trading regime. We will need to establish a reliable system of monitoring and verification to ensure that everyone plays by the rules. But that's the case with almost all international agreements, from arms control to intellectual property rights.

Another key piece of our climate strategy is joint implementation. We propose that private-sector companies in developed countries be allowed to undertake emissions reductions projects in developing countries, and count these reductions against their own emissions. Joint implementation can harness the expertise and capital of the private sector to reduce global greenhouse gas emissions in a cost effective manner. In addition, developing countries reap substantial, long-term benefits from such a system, through the transfer of cutting-edge technologies and business practices. Moreover, as we have seen in Central America, joint implementation projects can provide an invaluable mechanism to protect forests and other critical habitats around the world.

The U.S. already has launched many successful demonstration projects of activities jointly implemented from forestry to energy conservation, from Costa Rica to the Czech Republic, from Belize to Bolivia to Russia.

Moving to the next point, we think that all developed nations should have to take significant emissions reductions measures. This brings me to the subject of targets and timetables for reducing emissions in developed countries.

President Clinton recognizes that the United States, as the world's leading emitter of greenhouse gases, must set a strong example for other nations around the world. That is why he announced on October 22 his proposal to stabilize US emissions of greenhouse gases at 1990 levels during the period of 2008–2012. This means that the emission average for that five year period cannot exceed 1990 levels. Given our projected rate of emissions growth of more than thirty percent, the President's proposal actually represents a significant decrease in our emissions. The President also pledged to reduce emissions below 1990 levels in the years after 2012. The President heartily endorsed market mechanisms such as the ones I have discussed as a way to achieve meaningful emissions reductions in a cost-effective manner.

Looking at the larger picture, we also need to recognize that action by the United States and the other industrialized nations will not, by itself, put the world on the road to stable greenhouse gas concentrations.

As I've said, climate change is a global issue, requiring a world-wide response. It's all one atmosphere, whether it's polluted by American power plants, Brazilian steel mills, or Korean traffic jams.

At present, developed country emissions account for approximately 60% of the global total. But developing country emissions are growing rapidly, and by 2020, will account for more than half of the world's emissions.

China, which is already the world's second largest emitter, will surpass the U.S. within 15 years.

Now, we do recognize that there are over one hundred developing nations, and they vary greatly in size and level of economic development. Each of those nations can and should take actions commensurate with its capabilities and responsibilities.

The U.S. proposal for developing country participation, therefore, has three elements:

First, we call on all nations, developed and developing, to advance the implementation of their existing commitments to undertake climate-friendly policies and measures.

All nations should increase their energy efficiency, eliminate subsidies, and emphasize market-oriented pricing; increase the use of renewable energies;

facilitate investment in climate-friendly technologies; and promote the development and sustainable management of forests and other carbon "sinks" and "reservoirs." These are all measures that are justified economically in their own right and can also help in solving other environmental problems.

Second, we ask that advanced developing countries, particularly those which have graduated to OECD [Organization of Economic Cooperation and Development] status, voluntarily undertake quantified emission limitations;

And third, we call for a new series of negotiations to develop quantified obligations for all countries, and to establish a "trigger" for the automatic application of these obligations, based upon agreed criteria.

To summarize, we regard the meaningful participation of developing countries as an essential part of a comprehensive Kyoto agreement, along with the legally-binding commitments for developed countries and the creation of cost-effective implementation mechanisms.

I want to note here that Kyoto is just a first step in a process that must be sustained over many more years.

Our work will not finish in Kyoto, but it is important that it begin with a serious and committed first step.

This is an ambitious, exciting, consuming agenda, not costless, not barrier-free, but doable. In fact, the question is no longer what to do, the question is how to facilitate what so clearly needs to be done. I believe that our legacy depends in large measure on our ability to understand and react to these new challenges. And science will continue to provide the underpinning for our response.

In 1948, when the notion of space exploration was still science fiction, the Astronomer Fred Hoyle said: "Once a photograph of the Earth, taken from the outside, is available, a new idea as powerful as any in history will be let loose."

Twenty years later, when space travel became a reality, the travelers themselves provided powerful testimony to Hoyle's sense of the unity of the world. Let me read to you from our own astronaut, James Irwin: "That beautiful, warm, living object looked so fragile, so delicate that if you touched it with a finger it would crumble and fall apart."

And now from a Russian cosmonaut: "After an orange cloud—formed as a result of a dust storm over the Sahara—reached the Philippines and settled there with rain, I understood that we are all sailing in the same boat."

In this last decade of the millennium, we have the power and enormous responsibility to captain that boat carefully. We also have the ability to shape change for the benefit of the entire world. The interests and intellectual capacity reflected in this room today bear a special burden in this regard. Working together, your talents, your energy and your power are more than the match for the challenges and the institutions involved.

The future habitability and stability of the world rests in your hands.

On November 2, 1999, Robert T. Watson, chair of the Intergovernmental Panel on Climate Change, addressed the United Nations Framework Convention on Climate Change. His statement, which follows, concerns the implementation of the Kyoto Protocol and discusses the potential adverse effects of a warming that is considered to be already underway.

Report to the Fifth Conference of the Parties of the United Nations Framework Convention on Climate Change

Mr. Chairman, excellencies, distinguished delegates, it is a pleasure and honor for me to address you today. The Intergovernmental Panel on Climate Change values its close collaboration with the Parties to the Framework Convention on Climate Change, its Secretariat and its subsidiary bodies, and prides itself on being responsive to addressing your needs. The current IPCC work program has been designed to provide the scientific and technical information that is needed to implement the Convention and the Kyoto Protocol.

As you debate the weighty issues associated with effective implementation of the Convention and the Kyoto Protocol let me remind you that it is not a question of whether the Earth's climate will change, but rather when, where and by how much. It is undisputed that the last decade has been the warmest this century, indeed the warmest for hundreds of years, and many parts of the world have suffered major heat-waves, floods, droughts and extreme weather events leading to significant economic losses and loss of life. While individual events cannot be directly linked to human-induced climate change, the frequency and magnitude of these types of events are expected to increase in a warmer world.

I would also like to remind you of some of the adverse impacts of changes in climate that are projected to occur in different parts of the world:

First, arid and semi-arid land areas in Africa, the Middle East and Southern Europe become even more water stressed than they are today;

Second, agricultural production in Africa and Latin America decreases;

Third, the incidence of vector-borne diseases, such as malaria, increases in tropical countries;

Fourth, tens of millions of people will be displaced by rising sea levels in Small Island States and low-lying deltaic areas; and

Fifth, major changes in the structure and functioning of critical ecological systems, particularly coral reefs and forests.

These adverse impacts will severely undermine the goal of sustainable development in many parts of the world, with developing countries, and the poor in developing countries, being most vulnerable.

In Kyoto, governments decided that actions were needed to limit greenhouse gas emissions from industrialized countries. It is clear that there are numerous domestic policies, practices and technologies in Annex I [developed] countries, and subject to further elaboration, the flexible market instruments, that can be used to reduce the net emissions of greenhouse gases in a wide range of economic sectors in a cost-effective manner. Now is the time for the ratification and implementation of the Kyoto Protocol if the commitment targets are to be realized.

While the Kyoto Protocol will not stabilize the atmospheric concentrations of greenhouse gases, it is widely recognized that it is an important first step towards achieving the goal of Article 2 of the Convention. However, realizing the ultimate goal of Article 2 will require, amongst other actions, enhanced public and private sector investments in energy research and development.

Mr. Chairman, distinguished delegates, I would like to bring to your attention what I believe is a potentially serious weakness in the approach that the international community is taking in addressing the global environmental issues of climate change, land degradation, loss of biodiversity, stratospheric ozone depletion, water resource degradation and forest loss. The basic problem is that the scientific and policy communities have tended to treat these "global" environmental issues in isolation. This is a fundamental mistake because they are highly coupled. For example, changes in the Earth's climate can significantly affect the structure, functioning and geographic boundaries of ecological systems, leading to changes in genetic, species and ecosystem diversity. Changes in the structure and functioning of ecological systems will in turn affect the Earth's climate by modifying the Earth's albedo and the biogeochemical cycling of a number of key greenhouse gases. Therefore, until these issues are addressed in a more integrated manner it will be difficult to formulate and implement an optimal set of policies, practices and technologies, and develop the most effective financing mechanisms.

Mr. Chairman, in the remainder of my presentation today I would like to briefly update you on the status on the work of the IPCC and on the potentially dire situation of the IPCC budget.

First, the Special Report on aviation and the global atmosphere has been approved. One of the primary conclusions was that aircraft emit gases and particles directly into the upper troposphere and lower stratosphere where they have an impact on atmospheric composition and contribute to climate change. The best estimate of the radiative forcing by aircraft in 1992—a measure of climate change—is about 3.5% of the total radiative forcing by all human activities, projected to rise to between 3.5 and 15%, with a best estimate of 5%, by 2050 relative to the mid-range IPCC IS92a [mid-range emissions] scenario. Further, the IPCC concluded that there is a range of options to reduce aviation emissions, including changes in aircraft and engine technology, fuel, operational practices, and regulatory and economic measures.

Second, the Special Report on technology transfer examines the flows of knowledge, experience and equipment among governments, private sector entities, financial institutions, NGOs [nongovernmental organizations], and research and education institutions, and the different roles that each of these stakeholders can play in facilitating the transfer of technologies to address climate change in the context of sustainable development. The draft Report concludes that the current efforts and established processes will not be sufficient to meet this challenge. It is clear that enhanced capacity is required in developing countries and that additional government actions can create the enabling environment for private sector technology transfers within and across national boundaries.

Third, the Special Report on emissions scenarios examines a wide range of plausible futures for greenhouse gas and aerosol precursor emissions over the next 100 years. The methodology for developing these new scenarios recognizes that there are interactions among the key determinants of population growth, economic growth, energy demand, energy prices and the level of research and development. The new scenarios do not include any additional climate policies but some do assume sulfur policies in a number of developing countries. The Report provisionally concludes that there is a very wide range of plausible emissions scenarios.

Fourth, the Special Report on land-use, land-use change, and forestry addresses a number of issues where the Parties to the Kyoto Protocol will need to make key decisions before the relevant Articles of the Protocol can be implemented, particularly with respect to definitions, the accounting system, a monitoring and reporting system, and inventory guidelines. In addition, the Report provides an assessment of the experience to date of land use, land use change and forestry projects (largely AIJ [Activities Implemented Jointly] projects), the future potential to reduce the net emissions of greenhouse gases through Articles 3.3, 3.4, 6 and 12, and a framework for evaluating the sustainable development implications of such activities.

Fifth, the philosophy and scope of the Third Assessment Report will:
—emphasize cross-sectoral issues, adaptation and the regional dimensions of climate change;
—place the issue of climate change more centrally within the concept of sustainable development; and
—identify the synergies and trade-offs between local, regional and global environmental issues, in particular the inter-linkages between climate change, biodiversity, water resources and land degradation.

Lastly. Mr. Chairman, I need to bring to the delegates' attention the dire financial situation of the IPCC. This arises because the IPCC is responding to the large number of SBSTA [Subsidiary Body for Scientific and Technological Advice] requests for Special Reports coincident with the preparation of the TAR [Third Assessment Report], and the increased participation of experts from developing countries. Unfortunately in spite of numerous requests for adequate budgetary support there are many OECD countries who are contributing little to nothing. This lack of financial commitment is rather disturbing given the incredible effort of the experts who give so freely of their time to assist the Parties to the UNFCCC [United Nations Framework Convention on Climate Change] and the Kyoto Protocol. It is even more remarkable given the numerous interventions by Parties applauding the work of the IPCC and requesting the IPCC to undertake even more studies to support the negotiations. If the IPCC is to continue to serve the needs of the Parties additional governments will have to contribute to the IPCC Trust Fund, and some of those who routinely contribute will have to increase their contributions. Additionally, the work program as defined by the GEF [Global Environment Facility] Council could be amended so that the GEF could be viewed as a source of potential funding. This would appear to be quite appropriate given the emphasis that IPCC is placing on capacity building—a high priority of the Parties to the Convention and the Kyoto Protocol. I appeal to each government representative at this meeting to help resolve this untenable budgetary situation.

In conclusion, I would like to thank you for allowing me to address you today and remind you that the climate of Planet Earth and the welfare of future generations is in your hands.

The following statement is the testimony of Jerry D. Mahlman, director of the National Oceanic and Atmospheric Administration's Geophysical Fluid Dynamics Laboratory, to the U.S. Senate Committee on Commerce, Science, and Transportation. In this testimony, given on May 17, 2000, Mahlman offers a scientific perspective of the likelihood of an increased greenhouse warming, even given the current uncertainties in climate modeling.

Uncertainties In Climate Change Modeling

Mr. Chairman:

My name is Jerry Mahlman. I am the Director of the Geophysical Fluid Dynamics Laboratory of NOAA. For over thirty years our Laboratory has been a world leader in modeling the earth's climate. I will evaluate scientific projections of climate change as well as their current uncertainties.

We have long known that buildups of atmospheric carbon dioxide and other gases have the potential to warm earth's climate, through the so-called "greenhouse" effect. Today, I will discuss modeling the projections of climate changes due to these increasing greenhouse gases for a time around the middle of the century.

Because I speak with credentials as a physical scientist, I do not offer personal opinions on what society should do about these projected climate changes. Societal actions in response to greenhouse warming involve value and policy judgements that are beyond the realm of climate science.

At the onset, please recognize that a major international effort to assess climate warming was completed in 1996. This is "The Intergovernmental Panel on Climate Change Assessment" (IPCC). The IPCC was established in 1988 by the United Nations Environment Programme and the World Meteorological Organization to assess the available information on climate change and its environmental and economic impacts. This was the most widely accepted assessment ever on climate change. The 2001 IPCC Assessment will be completed soon. I expect only small changes in its major conclusions, mainly concerning some important increases in scientific confidence.

I strongly recommend your use of the IPCC assessments as a foundation for your own evaluations. I also recommend their use as a point of departure for evaluating the credibility of opinions that disagree with them.

My information is derived from the strengths and weaknesses of climate models, climate theory, and widespread observations of the climate system. Climate models have improved in their ability to simulate the climate and its natural variability. Unfortunately, important uncertainties remain due to deficiencies in our scientific understanding and in computer power.

However, significant progress is expected over the next 10 years.

However, let me say at the outset: None of the uncertainties I will discuss can make current concerns about greenhouse warming go away. This problem is very real and will be with us for a very long time.

I will give my evaluation of current model predictions of climate change in the middle of the next century by setting simple "betting odds." By

"Virtually Certain," I mean that there is no plausible alternative; in effect, the bet is off the books. "Very Probable" means I estimate about a 9 out of 10 chance that this will happen within the range projected; "Probable" implies about a 2 out of 3 chance. "Uncertain" means a plausible effect, but which lacks appropriate evidence. Essentially, I set the odds; you choose your bet. My analysis is presented in decreasing levels of confidence.

- Human-Caused Increasing Greenhouse Gases (virtually certain). There is no remaining doubt that increasing greenhouse gases are due to human activities.
- Radiative Effect of Increased Greenhouse Gases (virtually certain). Greenhouse gases absorb and reradiate infrared radiation. Independent of other factors, this property acts to produce an increased heating effect on the planet.
- A Doubling of Carbon Dioxide Expected (virtually certain). Atmospheric carbon dioxide amounts are expected to double over pre-industrial levels in this century. Current emissions growth is on track to quadruple atmospheric carbon dioxide.
- Long Time to Draw Down Excess Carbon Dioxide (virtually certain). We know that it takes decades to centuries to produce a large buildup of greenhouse gases. Much less appreciated is that a "return to normal" from high carbon dioxide levels would require many additional centuries.
- Global Surface Warming over the Past Century (virtually certain). The measured 20th century warming in the surface temperature records of over one degree Fahrenheit is undoubtedly real. Its cause is very probably due mostly to added greenhouse gases. No other hypothesis is nearly as credible.
- Future Global-Mean Surface Warming (very probable). For the middle of the next century, global-mean surface warming is estimated to be in the range of 2 to 6 Fahrenheit, with continued increases for the rest of the century. The largest uncertainty is due to the effects of clouds.
- Increased Summertime Heat Index (very probable). In warm, moist subtropical climates the summertime heat index effect is expected to magnify the warming impact felt by humans by an additional 50%.
- Rise in Global Mean Sea Level (very probable). A further rise of 4–12 inches in mean sea level by the year 2050 is estimated due to thermal expansion of warmer sea water. Continued sea level rise is expected for many centuries, probably to much higher levels.
- Summer Mid-Continental Dryness and Warming (probable). Model studies predict a marked decrease of soil moisture over summer mid-

latitude continents. This projection remains sensitive to model assumptions.
- Increased Tropical Storm Intensities (probable). A warmer, wetter atmosphere will likely lead to increased intensities of tropical storms, such as hurricanes. We still know little about changes in the number of hurricanes.
- Increased Numbers of Weather Disturbances (uncertain). Although many speak of more large-scale storms, there is still no solid evidence for this.
- Global and Regional Details of the Next 25 Years (uncertain).
  The predicted warming up to now is not yet large compared to natural climate fluctuations. On these shorter time scales, the natural fluctuations can artificially reduce or enhance apparent measured greenhouse warming signals, especially so on regional scales.

Even though these uncertainties are daunting, important advances have already been achieved in observing, understanding, and modeling the climate. Today's models can simulate many aspects of climate and its changes. Although major progress has been made, much more needs to be learned. More efforts are needed world-wide to provide a long-term climate measuring system.

Focussed research into climate processes must be continued. Theories must be formulated and re-evaluated in the light of newer data. Climate modeling efforts must receive resources that are in balance with the broader scientific programs.

The U.S. Global Change Research Program has already made important progress on these fronts. However, patient, sustained efforts will be required in the years ahead. Through long-term research and measurements, uncertainties will decrease and confidence for predicting climate changes will increase.

In summary, the greenhouse warming effect is quite real. The state of the science is strong, but important uncertainties remain.

Finally, it is a "virtually certain" bet that this problem will refuse to go away, no matter what is said or done about it over the next five years.

Thank you, Mr. Chairman. That concludes my testimony.

The following selection is from the Intergovernmental Panel on Climate Change (IPCC) assessment, referred to by both Watson and Mahlman in the previous selections. It provides an overview of the anthropogenic increase of various greenhouse gases and discusses the likely response of the planet to a corresponding warming of the earth's surface. Hundreds of authors from countries worldwide con-

Excerpts from IPCC Second Assessment Synthesis of Scientific-Technical Information Relevant to Interpreting Article 2 of the UN Framework Convention on Climate Change

ANTHROPOGENIC INTERFERENCE WITH THE CLIMATE SYSTEM

Interference to the Present Day

2.1 In order to understand what constitutes concentrations of greenhouse gases that would present dangerous interference with the climate system, it is first necessary to understand current atmospheric concentrations and trends of greenhouse gases, and their consequences (both present and projected) to the climate system.

2.2 The atmospheric concentrations of the greenhouse gases, and among them, carbon dioxide (CO_2), methane (CH_4) and nitrous oxide (N_2O), have grown significantly since preindustrial times (about 1750 A.D.): CO_2 from about 280 to almost 360 ppmv, CH_4 from 700 to 1720 ppbv and N_2O from about 275 to about 310 ppbv. These trends can be attributed largely to human activities, mostly fossil-fuel use, land-use change and agriculture. Concentrations of other anthropogenic greenhouse gases have also increased. An increase of greenhouse gas concentrations leads on average to an additional warming of the atmosphere and the Earth's surface. Many greenhouse gases remain in the atmosphere—and affect climate—for a long time.

2.3 Tropospheric aerosols resulting from combustion of fossil fuels, biomass burning and other sources have led to a negative direct forcing and possibly also to a negative indirect forcing of a similar magnitude. While the negative forcing is focused in particular regions and subcontinental areas, it can have continental to hemispheric scale effects on climate patterns. Locally, the aerosol forcing can be large enough to more than offset the positive forcing due to greenhouse gases. In contrast to the long-lived greenhouse gases, anthropogenic aerosols are very short-lived in the atmosphere and hence their radiative forcing adjusts rapidly to increases or decreases in emissions.

2.4 Global mean surface temperature has increased by between about 0.3 and 0.6°C since the late 19th century, a change that is unlikely to be entirely natural in origin. The balance of evidence, from changes in global mean surface air temperature and from changes in geographical, seasonal and vertical patterns of atmospheric temperature, suggests a discernible human influence on global climate. There are uncertainties in key factors, including the magnitude and patterns of long-term natural variability. Global sea level has risen by between 10 and 25 cm over the past 100 years and much of the rise may be related to the increase in global mean temperature.

2.5   There are inadequate data to determine whether consistent global changes in climate variability or weather extremes have occurred over the 20th century. On regional scales there is clear evidence of changes in some extremes and climate variability indicators. Some of these changes have been toward greater variability, some have been toward lower variability. However, to date it has not been possible to firmly establish a clear connection between these regional changes and human activities.

Possible Consequences of Future Interference

2.6   In the absence of mitigation policies or significant technological advances that reduce emissions and/or enhance sinks, concentrations of greenhouse gases and aerosols are expected to grow throughout the next century. The IPCC has developed a range of scenarios, IS92a–f, of future greenhouse gas and aerosol precursor emissions based on assumptions concerning population and economic growth, land-use, technological changes, energy availability and fuel mix during the period 1990 to 2100. By the year 2100, carbon dioxide emissions under these scenarios are projected to be in the range of about 6 GtC [Gigatons Carbon] per year, roughly equal to current emissions, to as much as 36 GtC per year, with the lower end of the IPCC range assuming low population and economic growth to 2100. Methane emissions are projected to be in the range 540 to 1170 Tg [teragrams] CH_4 per year (1990 emissions were about 500 Tg CH_4); nitrous oxide emissions are projected to be in the range 14 to 19 Tg N per year (1990 emissions were about 13 Tg N). In all cases, the atmospheric concentrations of greenhouse gases and total radiative forcing continue to increase throughout the simulation period of 1990 to 2100.

2.7   For the mid-range IPCC emission scenario, IS92a, assuming the "best estimate" value of climate sensitivity and including the effects of future increases in aerosol concentrations, models project an increase in global mean surface temperature relative to 1990 of about 2°C by 2100. This estimate is approximately one-third lower than the "best estimate" in 1990. This is due primarily to lower emission scenarios (particularly for CO_2 and CFCs), the inclusion of the cooling effect of sulphate aerosols, and improvements in the treatment of the carbon cycle. Combining the lowest IPCC emission scenario (IS92c) with a "low" value of climate sensitivity and including the effects of future changes in aerosol concentrations leads to a projected increase of about 1°C by 2100. The corresponding projection for the highest IPCC scenario (IS92e) combined with a "high" value of climate sensitivity gives a warming of about 3.5°C. In all cases the average rate of warming would probably be greater than any seen in the last 10,000 years, but the actual annual to decadal changes would include considerable natural variability. Regional temperature changes could differ substantially from the global mean value. Because of the thermal inertia of the oceans, only 50–90% of the eventual equilibrium temperature change would have been realized by 2100 and temperature would continue to increase beyond 2100, even if concentrations of greenhouse gases were stabilized by that time.

2.8 Average sea level is expected to rise as a result of thermal expansion of the oceans and melting of glaciers and ice-sheets. For the IS92a scenario, assuming the "best estimate" values of climate sensitivity and of ice melt sensitivity to warming, and including the effects of future changes in aerosol concentrations, models project an increase in sea level of about 50 cm from the present to 2100. This estimate is approximately 25% lower than the "best estimate" in 1990 due to the lower temperature projection, but also reflecting improvements in the climate and ice melt models. Combining the lowest emission scenario (IS92c) with the "low" climate and ice melt sensitivities and including aerosol effects gives a projected sea-level rise of about 15 cm from the present to 2100. The corresponding projection for the highest emission scenario (IS92e) combined with "high" climate and ice-melt sensitivities gives a sea-level rise of about 95 cm from the present to 2100. Sea level would continue to rise at a similar rate in future centuries beyond 2100, even if concentrations of greenhouse gases were stabilized by that time, and would continue to do so even beyond the time of stabilization of global mean temperature. Regional sea-level changes may differ from the global mean value owing to land movement and ocean current changes.

2.9 Confidence is higher in the hemispheric-to-continental scale projections of coupled atmosphere-ocean climate models than in the regional projections, where confidence remains low. There is more confidence in temperature projections than hydrological changes.

2.10 All model simulations, whether they were forced with increased concentrations of greenhouse gases and aerosols or with increased concentrations of greenhouse gases alone, show the following features: greater surface warming of the land than of the sea in winter; a maximum surface warming in high northern latitudes in winter, little surface warming over the Arctic in summer; an enhanced global mean hydrological cycle, and increased precipitation and soil moisture in high latitudes in winter. All these changes are associated with identifiable physical mechanisms.

2.11 Warmer temperatures will lead to a more vigorous hydrological cycle; this translates into prospects for more severe droughts and/or floods in some places and less severe droughts and/or floods in other places. Several models indicate an increase in precipitation intensity, suggesting a possibility for more extreme rainfall events. Knowledge is currently insufficient to say whether there will be any changes in the occurrence or geographical distribution of severe storms, e.g., tropical cyclones.

2.12 There are many uncertainties and many factors currently limit our ability to project and detect future climate change. Future unexpected, large and rapid climate system changes (as have occurred in the past) are, by their nature, difficult to predict. This implies that future climate changes may also involve "surprises". In particular, these arise from the non-linear nature of the climate system. When rapidly forced, non-linear systems are especially subject to unexpected behaviour. Progress can be made by investigating non-linear processes and sub-components of the climatic system. Examples of such non-linear behaviour include rapid circulation changes in the North Atlantic and feedbacks associated with terrestrial ecosystem changes.

SENSITIVITY AND ADAPTATION OF SYSTEMS TO CLIMATE CHANGE

3.1 This section provides scientific and technical information that can be used, inter alia, in evaluating whether the projected range of plausible impacts constitutes "dangerous anthropogenic interference with the climate system" as referred to in Article 2, and in evaluating adaptation options. However, it is not yet possible to link particular impacts with specific atmospheric concentrations of greenhouse gases.

3.2 Human health, terrestrial and aquatic ecological systems, and socio-economic systems (e.g., agriculture, forestry, fisheries and water resources) are all vital to human development and well-being and are all sensitive to both the magnitude and the rate of climate change. Whereas many regions are likely to experience the adverse effects of climate change—some of which are potentially irreversible—some effects of climate change are likely to be beneficial. Hence, different segments of society can expect to confront a variety of changes and the need to adapt to them.

3.3 Human-induced climate change represents an important additional stress, particularly to the many ecological and socio-economic systems already affected by pollution, increasing resource demands, and non-sustainable management practices. The vulnerability of human health and socioeconomic systems—and, to a lesser extent, ecological systems—depends upon economic circumstances and institutional infrastructure. This implies that systems typically are more vulnerable in developing countries where economic and institutional circumstances are less favourable.

3.4 Although our knowledge has increased significantly during the last decade and qualitative estimates can be developed, quantitative projections of the impacts of climate change on any particular system at any particular location are difficult because regional-scale climate change projections are uncertain; our current understanding of many critical processes is limited; systems are subject to multiple climatic and non-climatic stresses, the interactions of which are not always linear or additive; and very few studies have considered dynamic responses to steadily increasing concentrations of greenhouse gases or the consequences of increases beyond a doubling of equivalent atmospheric CO_2 concentrations.

3.5 Unambiguous detection of climate-induced changes in most ecological and social systems will prove extremely difficult in the coming decades. This is because of the complexity of these systems, their many non-linear feedbacks, and their sensitivity to a large number of climatic and non-climatic factors, all of which are expected to continue to change simultaneously. As future climate extends beyond the boundaries of empirical knowledge (i.e., the documented impacts of climate variation in the past), it becomes more likely that actual outcomes will include surprises and unanticipated rapid changes.

Sensitivity of Systems

Terrestrial and Aquatic Ecosystems

3.6 Ecosystems contain the Earth's entire reservoir of genetic and species diversity and provide many goods and services including: (i) providing food, fibre, medicines and energy; (ii) processing and storing carbon and other nutrients; (iii) assimilating wastes, purifying water, regulating water runoff, and controlling floods, soil degradation and beach erosion; and (iv) providing opportunities for recreation and tourism. The composition and geographic distribution of many ecosystems (e.g., forests, rangelands, deserts, mountain systems, lakes, wetlands and oceans) will shift as individual species respond to changes in climate; there will likely be reductions in biological diversity and in the goods and services that ecosystems provide society. Some ecological systems may not reach a new equilibrium for several centuries after the climate achieves a new balance. This section illustrates the impact of climate change on a number of selected ecological systems.

3.7 Forests: Models project that as a consequence of possible changes in temperature and water availability under doubled equivalent CO_2 equilibrium conditions, a substantial fraction (a global average of one-third, varying by region from one-seventh to two-thirds) of the existing forested area of the world will undergo major changes in broad vegetation types—with the greatest changes occurring in high latitudes and the least in the tropics. Climate change is expected to occur at a rapid rate relative to the speed at which forest species grow, reproduce and re-establish themselves. Therefore, the species composition of forests is likely to change; entire forest types may disappear, while new assemblages of species and hence new ecosystems may be established. Large amounts of carbon could be released into the atmosphere during transitions from one forest type to another because the rate at which carbon can be lost during times of high forest mortality is greater than the rate at which it can be gained through growth to maturity.

3.8 Deserts and desertification: Deserts are likely to become more extreme—in that, with few exceptions, they are projected to become hotter but not significantly wetter. Temperature increases could be a threat to organisms that exist near their heat tolerance limits. Desertification—land degradation in arid, semi-arid and dry sub-humid areas resulting from various factors, including climatic variations and human activities—is more likely to become irreversible if the environment becomes drier and the soil becomes further degraded through erosion and compaction.

3.9 Mountain ecosystems: The altitudinal distribution of vegetation is projected to shift to higher elevation; some species with climatic ranges limited to mountain tops could become extinct because of disappearance of habitat or reduced migration potential.

3.10 Aquatic and coastal ecosystems: In lakes and streams, warming would have the greatest biological effects at high latitudes, where biological productivity would increase, and at the low-latitude boundaries of cold- and cool-water species ranges, where extinctions would be greatest. The geographical distribution of wetlands is likely to shift with changes in temperature and precipitation. Coastal systems are economically and ecologically important and are expected to vary widely in their response to changes in climate and sea level. Some coastal ecosystems are particularly at risk, including saltwater marshes, mangrove ecosystems, coastal wetlands, sandy beaches, coral reefs, coral atolls and river deltas. Changes in these ecosystems would have major negative effects on tourism, freshwater supplies, fisheries and biodiversity.

Hydrology and Water Resources Management

3.11 Models project that between one-third and one-half of existing mountain glacier mass could disappear over the next hundred years. The reduced extent of glaciers and depth of snow cover also would affect the seasonal distribution of river flow and water supply for hydroelectric generation and agriculture. Anticipated hydrological changes and reductions in the areal extent and depth of permafrost could lead to large-scale damage to infrastructure, an additional flux of carbon dioxide into the atmosphere, and changes in processes that contribute to the flux of methane into the atmosphere.

3.12 Climate change will lead to an intensification of the global hydrological cycle and can have major impacts on regional water resources. Changes in the total amount of precipitation and in its frequency and intensity directly affect the magnitude and timing of runoff and the intensity of floods and droughts; however, at present, specific regional effects are uncertain. Relatively small changes in temperature and precipitation, together with the non-linear effects on evapotranspiration and soil moisture, can result in relatively large changes in runoff, especially in arid and semi-arid regions. The quantity and quality of water supplies already are serious problems today in many regions, including some low-lying coastal areas, deltas and small islands, making countries in these regions particularly vulnerable to any additional reduction in indigenous water supplies.

Agriculture and Forestry

3.13 Crop yields and changes in productivity due to climate change will vary considerably across regions and among localities, thus changing the patterns of production. Productivity is projected to increase in some areas and decrease in others, especially the tropics and subtropics. Existing studies show that on the whole, global agricultural production could be maintained relative to baseline production in the face of climate change projected under doubled equivalent CO_2 equilibrium conditions. This conclusion takes into account the beneficial effects of CO_2 fertilization but does not allow for changes in agricultural pests and the possible effects of changing climatic variability. However, focusing on global agricultural production does not address the potentially serious consequences of large differences at local and regional scales, even at mid-latitudes. There may be increased risk of hunger and famine in some locations; many of the world's poorest people—particularly those living in subtropical and tropical areas and dependent on isolated agricultural systems in semi-arid and arid regions—are most at risk of increased hunger. Global wood supplies during the next century may become increasingly inadequate to meet projected consumption due to both climatic and non-climatic factors.

Human Infrastructure

3.14 Climate change clearly will increase the vulnerability of some coastal populations to flooding and erosional land loss. Estimates put about 46 million people per year currently at risk of flooding due to storm surges.

In the absence of adaptation measures, and not taking into account anticipated population growth, 50-cm sea-level rise would increase this number to about 92 million; a 1-meter sea-level rise would raise it to about 118 million. Studies using a 1-meter projection show a particular risk for small islands and deltas. This increase is at the top range of IPCC Working Group I estimates for 2100; it should be noted, however, that sea level is actually projected to continue to rise in future centuries beyond 2100. Estimated land losses range from 0.05% in Uruguay, 1.0% for Egypt, 6% for the Netherlands and 17.5% for Bangladesh to about 80% for the Majuro Atoll in the Marshall Islands, given the present state of protection systems. Some small island nations and other countries will confront greater vulnerability because their existing sea and coastal defense systems are less well established. Countries with higher population densities would be more vulnerable. Storm surges and flooding could threaten entire cultures. For these countries, sea-level rise could force internal or international migration of populations.

Human health

3.15 Climate change is likely to have wide-ranging and mostly adverse impacts on human health, with significant loss of life. Direct health effects include increases in (predominantly cardio-respiratory) mortality and illness due to an anticipated increase in the intensity and duration of heat waves. Temperature increases in colder regions should result in fewer cold-related deaths. Indirect effects of climate change, which are expected to predominate, include increases in the potential transmission of vector-borne infectious diseases (e.g., malaria, dengue, yellow fever and some viral encephalitis) resulting from extensions of the geographical range and season for vector organisms. Models (that entail necessary simplifying assumptions) project that temperature increases of 3–5°C (compared to the IPCC projection of 1–3.5°C by 2100) could lead to potential increases in malaria incidence (of the order of 50–80 million additional annual cases, relative to an assumed global background total of 500 million cases), primarily in tropical, subtropical and less well-protected temperate-zone populations. Some increases in non-vector-borne infectious diseases—such as salmonellosis, cholera and giardiasis—also could occur as a result of elevated temperatures and increased flooding. Limitations on freshwater supplies and on nutritious food, as well as the aggravation of air pollution, will also have human health consequences.

3.16 Quantifying the projected impacts is difficult because the extent of climate-induced health disorders depends on numerous coexistent and interacting factors that characterize the vulnerability of the particular population, including environmental and socio-economic circumstances, nutritional and immune status, population density and access to quality health care services. Hence, populations with different levels of natural, technical and social resources would differ in their vulnerability to climate-induced health impacts.

Technology and Policy Options for Adaptation

3.17 Technological advances generally have increased adaptation options for managed systems. Adaptation options for freshwater resources include more efficient management of existing supplies and infrastructure; institutional arrangements to limit future demands/promote conservation; improved monitoring and forecasting systems for floods/droughts; rehabilitation of watersheds, especially in the tropics; and construction of new reservoir capacity. Adaptation options for agriculture—such as changes in types and varieties of crops, improved water-management and irrigation systems, and changes in planting schedules and tillage practices—will be important in limiting negative effects and taking advantage of beneficial changes in climate. Effective coastal-zone management and land-use planning can help direct population shifts away from vulnerable locations such as flood plains, steep hillsides and low-lying coastlines. Adaptive options to reduce health impacts include protective technology (e.g., housing, air conditioning, water purification and vaccination), disaster preparedness and appropriate health care.

3.18 However, many regions of the world currently have limited access to these technologies and appropriate information. For some island nations, the high cost of providing adequate protection would make it essentially infeasible, especially given the limited availability of capital for investment. The efficacy and cost-effective use of adaptation strategies will depend upon the availability of financial resources, technology transfer, and cultural, educational, managerial, institutional, legal and regulatory practices, both domestic and international in scope. Incorporating climate-change concerns into resource-use and development decisions and plans for regularly scheduled investments in infrastructure will facilitate adaptation.

OZONE DEPLETION

Another topic to receive much international attention and debate in recent years is the loss of ozone in the stratosphere, particularly in the polar regions. The World Meteorological Society released its most recent assessment of ozone depletion in 1998. Excerpts from the executive summary, including implications for policy formulation, are reprinted here.

Scientific Assessment of Ozone Depletion: From the 1998 Assessment Executive Summary

RECENT MAJOR SCIENTIFIC FINDINGS AND OBSERVATIONS

Since the Scientific Assessment of Ozone Depletion: 1994, significant advances have continued to be made in the understanding of the impact of human activities on the ozone layer, the influence of changes in chemical composition on the radiative balance of the Earth's climate, and, indeed, the coupling of the ozone layer and the climate system. Numerous laboratory investigations, atmospheric observations, and theoretical and modeling studies have produced several key ozone- and climate-related findings:

The total combined abundance of ozone-depleting compounds in the lower atmosphere peaked in about 1994 and is now slowly declining. Total chlorine is declining, but total bromine is still increasing. As forecast in the 1994 Assessment, the long period of increasing total chlorine abundances—primarily from the chlorofluorocarbons (CFCs), carbon tetrachloride (CCl_4), and methyl chloroform (CH_3CCl_3)—has ended. The peak total tropospheric chlorine abundance was 3.7 ± 0.1 parts per billion (ppb) between mid-1992 and mid-1994. The declining abundance of total chlorine is due principally to reduced emissions of methyl chloroform. Chlorine from the major CFCs is still increasing slightly. The abundances of most of the halons continue to increase (for example, Halon-1211, almost 6% per year in 1996), but the rate has slowed in recent years. These halon increases are likely to be due to emissions in the 1990s from the halon "bank," largely in developed countries, and new production of halons in developing countries. The observed abundances of CFCs and chlorocarbons in the lower atmosphere are consistent with reported emissions.

The observed abundances of the substitutes for the CFCs are increasing. The abundances of the hydrochlorofluorocarbons (HCFCs) and hydrofluorocarbons (HFCs) are increasing as a result of a continuation of earlier uses and of their use as substitutes for the CFCs. In 1996, the HCFCs contributed about 5% to the tropospheric chlorine from the long-lived gases. This addition from the substitutes offsets some of the decline in tropospheric chlorine associated with methyl chloroform, but is nevertheless about 10 times less than that from the total tropospheric chlorine growth rate throughout the 1980s. The atmospheric abundances of HCFC-141b and

HCFC-142b calculated from reported emissions data are factors of 1.3 and 2, respectively, smaller than observations. Observed and calculated abundances agree for HCFC-22 and HFC-134a.

The combined abundance of stratospheric chlorine and bromine is expected to peak before the year 2000. The delay in this peak in the stratosphere compared with the lower atmosphere reflects the average time required for surface emissions to reach the lower stratosphere. The observations of key chlorine compounds in the stratosphere up through the present show the expected slower rate of increase and show that the peak had not occurred at the time of the most recent observations that were analyzed for this Assessment.

The role of methyl bromide as an ozone-depleting compound is now considered to be less than was estimated in the 1994 Assessment, although significant uncertainties remain. The current best estimate of the Ozone Depletion Potential (ODP) for methyl bromide (CH_3Br) is 0.4, compared with an ODP of 0.6 estimated in the previous Assessment. The change is due primarily to both an increase in the estimate of ocean removal processes and the identification of an uptake by soils, with a smaller contribution from the change in our estimate of the atmospheric removal rate. Recent research has shown that the science of atmospheric methyl bromide is complex and still not well understood. The current understanding of the sources and sinks of atmospheric methyl bromide is incomplete.

The rate of decline in stratospheric ozone at midlatitudes has slowed; hence, the projections of ozone loss made in the 1994 Assessment are larger than what has actually occurred. Total column ozone decreased significantly at midlatitudes (25–60°) between 1979 and 1991, with estimated linear downward trends of 4.0, 1.8, and 3.8% per decade, respectively, for northern midlatitudes in winter/spring, northern midlatitudes in summer/fall, and southern midlatitudes year round. However, since 1991 the linear trend observed during the 1980s has not continued, but rather total column ozone has been almost constant at midlatitudes in both hemispheres since the recovery from the 1991 Mt. Pinatubo eruption. The observed total column ozone losses from 1979 to the period 1994–1997 are about 5.4, 2.8, and 5.0%, respectively, for northern midlatitudes in winter/spring, northern midlatitudes in summer/fall, and southern midlatitudes year round, rather than the values projected in the 1994 Assessment assuming a linear trend: 7.6, 3.4, and 7.2%, respectively. The understanding of how changes in stratospheric chlorine/bromine and aerosol loading affect ozone suggests some of the reasons for the unsuitability of using a linear extrapolation of the pre-1991 ozone trend to the present.

The link between the long-term build-up of chlorine and the decline of ozone in the upper stratosphere has been firmly established. Model predictions based on the observed build-up of stratospheric chlorine in the upper stratosphere indicate a depletion of ozone that is in good quantitative agreement with the altitude and latitude dependence of the measured ozone decline during the past several decades, which peaks at about 7% per decade near 40 km at midlatitudes in both hemispheres.

The springtime Antarctic ozone hole continues unabated. The extent of ozone depletion has remained essentially unchanged since the early 1990s. This behavior is expected given the near-complete destruction of ozone within the Antarctic lower stratosphere during springtime. The factors contributing to the continuing depletion are well understood.

The late-winter/spring ozone values in the Arctic were unusually low in 6 out of the last 9 years, the 6 being years that are characterized by unusually cold and protracted stratospheric winters. The possibility of such depletions was predicted in the 1989 Assessment. Minimum Arctic vortex temperatures are near the threshold for large chlorine activation. Therefore, the year-to-year variability in temperature, which is driven by meteorology, leads to particularly large variability in ozone for current chlorine loading. As a result, it is not possible to forecast the behavior of Arctic ozone for a particular year. Elevated stratospheric halogen abundances over the next decade or so imply that the Arctic will continue to be vulnerable to large ozone losses.

The understanding of the relation between increasing surface UV-B radiation and decreasing column ozone has been further strengthened by ground-based observations, and newly developed satellite methods show promise for establishing global trends in UV radiation. The inverse dependence of surface UV radiation and the overhead amount of ozone, which was demonstrated in earlier Assessments, has been further demonstrated and quantified by ground-based measurements under a wide range of atmospheric conditions. In addition, the influences of other variables, such as clouds, particles, and surface reflectivity, are better understood. These data have assisted the development of a satellite-based method to estimate global UV changes, taking into account the role of cloud cover. The satellite estimates for 1979–1992 indicate that the largest UV increases occur during spring at high latitudes in both hemispheres.

Stratospheric ozone losses have caused a cooling of the global lower stratosphere and global-average negative radiative forcing of the climate system. The decadal temperature trends in the stratosphere have now been better quantified. Model simulations indicate that much of the observed downward trend in lower stratospheric temperatures (about 0.6°C per decade over 1979–1994) is attributed to the ozone loss in the lower stratosphere. A lower stratosphere that is cooler results in less infrared radiation reaching the surface/troposphere system. Radiative calculations, using extrapolations based on the ozone trends reported in the 1994 Assessment for reference, indicate that stratospheric ozone losses since 1980 may have offset about 30% of the positive forcing due to increases in the well-mixed greenhouse gases (i.e., carbon dioxide, methane, nitrous oxide, and the halocarbons) over the same time period. The climatic impact of the slowing of midlatitude ozone trends and the enhanced ozone loss in the Arctic has not yet been assessed.

Based on past emissions of ozone-depleting substances and a projection of the maximum allowances under the Montreal Protocol into the future, the maximum ozone depletion is estimated to lie within the current decade or the next two decades, but its identification and the evidence for the recovery of the ozone layer lie still further ahead. The falloff of total chlorine and bromine abundances in the stratosphere in the next century will be much slower than the rate of increase observed in past decades, because of the slow rate at which natural processes remove these compounds from the stratosphere. The most vulnerable period for ozone depletion will be extended into the coming decades. However, extreme perturbations, such as natural events like volcanic eruptions, could enhance the loss from ozone-depleting chemicals. Detection of the beginning of the recovery of the ozone layer could be achievable early in the next century if decreasing chlorine and bromine abundances were the only factor. However, potential future increases or decreases in other gases important in ozone chemistry (such as nitrous oxide, methane, and water vapor) and climate change will influence the recovery of the ozone layer. When combined with the natural variability of the ozone layer, these factors imply that unambiguous detection of the beginning of the recovery of the ozone layer is expected to be well after the maximum stratospheric loading of ozone-depleting gases.

IMPLICATIONS FOR POLICY FORMULATION

The results from more than two decades of research have provided a progressively better understanding of the interaction of human activities and the chemistry and physics of the global atmosphere. New policy-relevant insights to the roles of trace atmospheric constituents have been conveyed to decision-makers through the international state-of-the-understanding assessment process. This information has served as a key input to policy decisions by governments, industry, and other organizations worldwide to limit the anthropogenic emissions of gases that cause environmental degradation: (1) the 1987 Montreal Protocol on ozone-depleting substances, and its subsequent Amendments and Adjustments, and (2) the 1997 Kyoto Protocol on substances that alter the radiative forcing of the climate system.

The research findings that are summarized above are of direct interest and significance as scientific input to governmental, industrial, and other policy decisions associated with the Montreal Protocol (ozone layer) and the Kyoto Protocol (climate change):

The Montreal Protocol is working. Global observations have shown that the combined abundance of anthropogenic chlorine-containing and bromine-containing ozone-depleting substances in the lower atmosphere peaked in 1994 and has now started to decline. One measure of success of the Montreal Protocol and its subsequent Amendments and Adjustments is the forecast of "the world that was avoided" by the Protocol: The abundance of ozone-depleting gases in 2050, the approximate time at which the ozone layer is now projected to recover to pre-1980 levels, would be at least 17 ppb of equivalent effective chlorine (this is based on the conservative assumption of a 3% per annum growth in ozone-depleting gases), which is about 5 times larger than today's value. Ozone depletion would be at least 50% at midlatitudes in the Northern Hemisphere and 70% at midlatitudes in the Southern Hemisphere, about 10 times larger than today. Surface UV-B radiation would at least double at midlatitudes in the Northern Hemisphere and quadruple at midlatitudes in the Southern Hemisphere compared with an unperturbed atmosphere. This compares to the current increases of 5% and 8% in the Northern and Southern Hemispheres, respectively, since 1980.

Furthermore, all of the above impacts would have continued to grow in the years beyond 2050. It is important to note that, while the provisions of the original Montreal Protocol in 1987 would have lowered the above growth rates, recovery (i.e., an improving situation) would have been impossible without the Amendments and Adjustments (London, 1990; Copenhagen, 1992; and Vienna, 1995).

The ozone layer is currently in its most vulnerable state. Total stratospheric loading of ozone-depleting substances is expected to maximize before the year 2000. All other things being equal, the current ozone losses (relative to the values observed in the 1970s) would be close to the maximum. These are:

about 6% at Northern Hemisphere midlatitudes in winter/spring;
about 3% at Northern Hemisphere midlatitudes in summer/fall;
about 5% at Southern Hemisphere midlatitudes on a year-round basis;
about 50% in the Antarctic spring; and
about 15% in the Arctic spring.

Such changes in ozone are predicted to be accompanied by increases in surface erythemal radiation of 7, 4, 6, 130, and 22%, respectively, if other influences such as clouds remain constant. It should be noted that these values for ozone depletion at midlatitudes are nearly a factor of 2 lower than projected in 1994, primarily because the linear trend in ozone observed in the 1980s did not continue in the 1990s. However, springtime depletion of ozone in Antarctica continues unabated at the same levels as observed in the early 1990s, and large depletions of ozone have been observed in the Arctic in most years since 1990, which are characterized by unusually cold and protracted winters.

Some natural and anthropogenic processes that do not in themselves cause ozone depletion can modulate the ozone loss from chlorine and bromine compounds, in some cases very strongly. For example, in coming decades midlatitude ozone depletion could be enhanced by major volcanic eruptions, and Arctic ozone depletion could be increased by cold polar temperatures, which in turn could be linked to greenhouse gases or to natural temperature fluctuations. On the other hand, increases in methane would tend to decrease chlorine-catalyzed ozone loss.

The current vulnerability to ozone depletion over the next few decades is primarily due to past use and emissions of the long-lived ozone-depleting substances. The options to reduce this vulnerability over the next two decades are thus rather limited. The main drivers of ozone change could be natural and anthropogenic processes not related to chlorine and bromine compounds, but to which the ozone layer is sensitive because of the elevated abundances of ozone-depleting substances.

The ozone layer will slowly recover over the next 50 years. The stratospheric abundance of halogenated ozone-depleting substances is expected to return to its pre-1980 (i.e., "unperturbed") level of 2 ppb chlorine equivalent by about 2050, assuming full compliance with the Montreal Protocol and its Amendments and Adjustments. The atmospheric abundances of global and Antarctic ozone will start to slowly recover within coming decades toward their pre-1980 levels once the stratospheric abundances of ozone-depleting (halogen) gases start to decrease. However, the future abundance of ozone will be controlled not only by the abundance of halogens, but also by the atmospheric abundances of methane, nitrous oxide, water vapor, and sulfate aerosols and by the Earth's climate. Therefore, for a given halogen loading in the future, atmospheric ozone abundance is unlikely to be the same as found in the past for the same halogen loading.

Few policy options are available to enhance the recovery of the ozone layer. Relative to the current, but not yet ratified, control measures (Montreal, 1997), the equivalent effective chlorine loading above the 1980 level, integrated from now until the 1980 level is re-attained, could be decreased by:

9% by eliminating global Halon-1211 emissions in the year 2000, thus requiring the complete elimination of all new production and destruction of all Halon-1211 in existing equipment;

7% by eliminating global Halon-1301 emissions in the year 2000, thus requiring the complete elimination of all new production and destruction of all Halon-1301 in existing equipment;

5% by eliminating the global production of all HCFCs in the year 2004;

2.5% by eliminating the global production of all CFCs and carbon tetrachloride in the year 2004;

1.6% by reducing the cap on HCFC production in developed countries from 2.8% to 2.0% in the year 2000, by advancing the phase-out from the year 2030 to 2015, and by instituting more rapid intermediate reductions; and

about 1% by eliminating the global production of methyl bromide beginning in 2004.

These policy actions would advance the date at which the abundance of effective chlorine returns to the 1980 value by 1–3 years. A complete and immediate global elimination of all emissions of ozone-depleting substances would result in the stratospheric halogen loading returning to the pre-1980 values by the year 2033. It should also be noted that if the currently allowed essential uses for metered dose inhalers are extended from the year 2000 to 2004, then the equivalent effective chlorine loading above the 1980 level would increase by 0.3%.

Failure to comply with the international agreements of the Montreal Protocol will affect the recovery of the ozone layer. For example, illegal production of 20–40 ktonnes per year of CFC-12 and CFC-113 for the next 10–20 years would increase the equivalent effective chlorine loading above the 1980 abundance, integrated from now until the 1980 abundance is reattained, by about 1–4% and delay the return to pre-1980 abundances by about a year.

The issues of ozone depletion and climate change are interconnected; hence, so are the Montreal and Kyoto Protocols. Changes in ozone affect the Earth's climate, and changes in climate and meteorological conditions affect the ozone layer, because the ozone depletion and climate change phenomena share a number of common physical and chemical processes. Hence, decisions taken (or not taken) under one Protocol have an impact on the aims of the other Protocol. For example, decisions made under the Kyoto Protocol with respect to methane, nitrous oxide, and carbon dioxide will affect the rate of recovery of ozone, while decisions regarding controlling HFCs may affect decisions regarding the ability to phase out ozone-depleting substances.

Chapter Nine

Professional Societies and Research Organizations

This chapter lists major meteorological and scientific organizations in the United States and internationally. Some of the organizations are professional societies dedicated to providing their members with information on meteorological and the atmospheric issues; others are organizations directly sponsoring atmospheric or meteorological research. The professional societies can be good starting places for information about meteorological careers and opportunities in both the private and public sectors. For persons interested in research, materials produced by the National Oceanic and Atmospheric Administration, the National Center for Atmospheric Research, the Environmental Protection Agency, and other research organizations provide excellent highlights of current and ongoing issues in the science.

American Association for the Advancement of Science
1200 New York Avenue, NW
Washington, DC 20005
(202) 326–6400
http://www.aaas.org

Founded in 1848, the AAAS is one of the oldest scientific societies in the United States. Membership is open to all individuals, and currently numbers over 143,000 individuals worldwide, including scientists, policymakers, educators, and private citizens from several countries. AAAS publishes the well-known weekly journal *Science*, which includes research reports, review articles, and discussions of public policy related to scientific and technological issues.

American Association of State Climatologists
c/o National Climatic Data Center
151 Patton Avenue
Asheville, NC 28801-5001
http://www.ncdc.noaa.gov/aasc.html
The American Association of State Climatologists is a professional scientific organization composed of state climatologists and directors of six regional climate centers in the United States. Persons interested in the goals and activities of the association, including assistant state climatologists, representatives of federal climate agences, retired state climatologists, or others interested in climate services, may serve as associate members. Membership in the association is about 150 and includes existing state climatologists in forty-seven states and Puerto Rico.

American Geophysical Union
2000 Florida Avenue, NW
Washington, DC 20009-1277
http://www.agu.org
The AGU is an international scientific society dedicated to advancing the understanding of the earth and its environment in space. The organization celebrated its seventy-fifth anniversary in 1998 and currently has 35,000 members. More than 30 percent of AGU members come from one of over 115 countries outside the United States. AGU publishes a number of scholarly journals, including *Geophysical Research Letters* and the *Journal of Geophysical Research*. A main component of AGU is the communication of scientific information to the public. To accomplish this goal, the organization sponsors media events, meetings, and other forums for discussing scientific issues.

American Meteorological Society
45 Beacon Street
Boston, MA 02108-3693
http://www.ametsoc.org/AMS
The AMS was created in 1919 as a scientific and professional organization serving the atmospheric and related sciences. With a current membership of over 11,000, the AMS works to develop and disseminate knowledge of the atmospheric and related oceanic and hydrologic sciences and to foster the advancement of professional applications in these areas. The AMS is a key source of educational information for students

and professionals in all facets of meteorology and atmospheric science. In addition to several monographs and other publications, the AMS produces nine journals, including the *Journal of Atmospheric Sciences, Journal of Applied Meteorology*, and *Journal of Climate*, as well as *Weather and Forecasting, Monthly Weather Review*, and the monthly *Bulletin of the American Meteorological Society*. The AMS offers a certification program for consulting meteorologists, as well as a seal of approval for professional weathercasters on radio and television. It also hosts twelve annual conferences covering such areas as atmospheric radiation, clouds and precipitation, hurricanes and tropical meteorology, and agricultural and forest meteorology.

Australian Meteorological and Oceanographic Society
P.O. Box 645E
Melbourne, VTC 3001 Australia
http://www.amos.org.au/

The AMOS is an independent society that supports and fosters interest in meteorology, oceanography, and other related sciences. It accomplishes this task by providing a forum for people with a common interest and publishing relevant material. The society helps all those with an interest in the environment, including research workers and professionals; people in industries affected by and affecting the atmosphere and oceans; and those who simply want to keep up with new findings. The AMOS provides support and fosters interest in meteorology and oceanography through its publications, including the quarterly *Australian Meteorological Magazine*, a scientific journal reporting the results of research in the atmospheric and related sciences, with an emphasis on the meteorology of the Australian region and the Southern Hemisphere. The society sponsors meetings, courses, grants, and prizes and represents the views of its members to government, institutes, and the public. The society keeps members informed on current news and activities through the regular distribution of the bimonthly *Bulletin of the Australian Meteorological and Oceanographic Society* and also publishes proceedings of international and local scientific meetings it has sponsored.

Canadian Meteorological and Oceanographic Society
112-150 Louis Pasteur
Ottawa, Ontario Canada K1N 6N5
(613) 562–5616

The CMOS was created in 1967 as the Canadian Meteorological Society and adopted its current name in 1977. The society comprises some 1100 members and subscribers, including students, corporations, institutions, and others who are involved in education, communications, the private sector, and the government. It addresses a broad range of national and international meteorological and oceanographic concerns, including weather and weather extremes, global warming, ozone depletion, and surface air quality. It also examines the effects of these concerns on all aspects of life in Canada, including forestry, agriculture, and fisheries. CMOS publishes the bimonthly *Bulletin*, the annual *Congress Program and Abstracts*, books, the *Annual Review*, and the internationally respected journal *Atmosphere-Ocean*.

Climate Specialty Group—Association of American Geographers
1710 Sixteenth Street, NW
Washington, DC 20009-3198
(202) 234-1450
http://www.geog.ucla.edu/csghome.html

The CSG is a subgroup of the Association of American Geographers. Its membership is focused on the following objectives: (1) to encourage climatological research; (2) to promote education in climatology; (3) to promote exchange of climate-related ideas and information; (4) to promote the interests of climatology with geography; and (5) to develop contacts and coordination with other climatological organizations.

Environmental Protection Agency
1200 Pennsylvania Avenue, NW
Washington, DC 20460
http://www.epa.gov/

The EPA was established on December 2, 1970, during Richard Nixon's term as U.S. president. Its mission is to protect human health and safeguard the natural environment—air, water, and land. The EPA protects Americans from significant health risks in the environment where they live, learn, and work. This goal is attained through national efforts to reduce environmental risk based on the best available scientific information, and through federal laws protecting human health and the environment. Through the EPA, the United States plays a leadership role in working with other nations to protect the global environment. The EPA has been successful in establishing over twenty-five years of environmental progress, including improvements to air and water quality.

Irish Meteorological Society
c/o Met Eirann
Glasnevin Hill
Dublin 9 Ireland
http://indigo.ie/~kcommins/metsocl.htm

The Irish Meteorological Society was founded in 1981 to promote an interest in meteorology and disseminate meteorological knowledge, both pure and applied. It includes members from all over the world who work in meteorology, aviation, agriculture, and marine concerns. It also includes teachers, lecturers, and anyone else who is interested in meteorology and the environment. The society publishes a quarterly members' newsletter and hosts several activities and lectures.

Meteorological Society of New Zealand
P.O. Box 6523
Te Aro
Wellington, New Zealand
http://metsoc.rsnz.govt.nz

The Meteorological Society of New Zealand was inaugurated in 1979 to encourage an interest in the atmosphere, weather, and climate, particularly as related to the New Zealand region. The society offers organized meetings, an annual professional journal (*Weather and Climate*), a quarterly newsletter, and a yearly conference on weather and climate issues. Membership includes meteorologists, climatologists, engineers, geographers, hydrologists, ecologists, economists, and others with an interest in weather and the society's objectives.

National Aeronautics and Space Administration
Headquarters Office
Washington, DC 20546
(202) 358-0000
http://www.nasa.gov/

NASA was established in 1958. In the years since, it has accomplished many great scientific and technological feats in air and space, as well as in the private sector. NASA remains a leading force in scientific research, especially in the area of earth science. The agency sponsors several research centers, including Langley Research Center in Hampton, Virginia; Ames Research Center in Moffett Field, California; the Goddard Space Flight Center in Greenbelt, Maryland; and the Goddard Institute for Space Studies in New York. NASA has pushed the frontier of atmospheric research with its earth-observing satellites, the most recent of which are part of the Earth Science Enterprise.

National Center for Atmospheric Research
1850 Table Mesa Drive
Boulder, CO 80303
(303) 497-1000
http://www.ncar.ucar.edu/ncar/index.html
The NCAR is sponsored by the National Science Foundation and managed cooperatively by the University Corporation for atmospheric research. Its mission includes providing state-of-the-art research tools and facilities to the entire atmospheric sciences community. It also facilitates the transfer of technology to public and private sectors. NCAR's personnel organize and conduct atmospheric and related research programs in collaboration with atmospheric research programs at universities. NCAR hosts a number of scientific divisions and programs, organized around specific functions related to various areas of research:

Atmospheric Chemistry Division
Advanced Study Program
Atmospheric Technology Division
Climate and Global Dynamics Division
Mesoscale and Microscale Meteorology Division
Environmental and Societal Impacts Group
High Altitude Observatory
Research Applications Program
Scientific Computing Division

National Council of Industrial Meteorologists
c/o Dr. George E. McVehil
44 Inverness Drive East, Building C
Englewood, CO 80112
The NCIM was founded in 1968 with the goal of furthering the development and practice of industrial meteorology. Members must be Certified Consulting Meteorologists as designated by the American Meteorological Society or possess equivalent qualifications. The members use their talents to solve a variety of meteorological problems encountered by a large group of industrial and governmental clients. As of 1997-1998, the NCIM had approximately sixty members from over twenty states.

National Oceanic and Atmospheric Administration
14th Street and Constitution Avenue, NW, Room 6013
Washington, DC 20230
(202) 482–6090
http://www.noaa.gov/index.html
NOAA was founded in 1970 with the goal of attaining a better understanding of the environment. This understanding would be paramount to monitoring and predicting the behavior of the atmosphere and oceans. NOAA, part of the Department of Commerce, comprises the National Ocean Service, the National Weather Service, the National Marine Fisheries Service, the National Environmental Satellite Data and Information Service, and the Office of Oceanic and Atmospheric Research. The Office of Oceanic and Atmospheric Research oversees twelve laboratories focused on various atmospheric and oceanic research issues:

- Aeronomy Laboratory
- Atlantic Oceanographic and Meteorological Laboratory
- Air Resources Laboratory
- Climate Diagnostics Laboratory
- Climate Monitoring and Diagnostics Laboratory
- Environmental Technology Laboratory
- Forecast Systems Laboratory
- Geophysical Fluid Dynamics Laboratory
- Great Lakes Environmental Research Laboratory
- National Severe Storms Laboratory
- Pacific Marine Environmental Laboratory
- Space Environment Center

National Science Foundation
4201 Wilson Boulevard
Arlington, VA 22230
(703) 306–1234
http://www.nsf.gov
The NSF is an independent agency of the U.S. government established by the National Science Foundation Act of 1950. Its mission is to promote the progress of science; advance national health, prosperity, and welfare; and secure the national defense. It accomplishes these activities through many mechanisms, including initiating and supporting grants for scientific and engineering research and programs, fostering the inter-

change of scientific information among scientists and engineers in the United States and foreign countries, and recommending and encouraging the pursuit of national policies for the promotion of basic research and education in the sciences and engineering.

National Weather Association
6704 Wolke Court
Montgomery, AL 36116–2134
http://www.nwas.org

The NWA was founded in 1975 to support and promote excellence in operational meteorology and related activities. With a current membership of over 2800 individual and 45 corporate participants, the NWA includes over 50 international members and subscribers. It provides grants to teachers to help improve the meteorology education for students in grades K–12 and also offers an NWA Seal of Approval to be earned by radio/TV weathercasters. Through the NWA, operational meteorologists, hydrologists, broadcasters, and individuals in related operational activities can share studies, techniques, news, and more. The association produces a monthly newsletter and hosts an annual meeting.

Royal Meteorological Society
104 Oxford Road
Reading, Berkshire RG1 7LL UK
http://www.royal-met-soc.org/uk/

The Royal Meteorological Society was founded under a royal charter for the advancement of meteorological science. The advancement of meteorological science is accomplished through publishing the results of new research and the greater spread of well-founded knowledge. The society serves not only research workers and professional meteorologists but also persons whose work is affected by the weather or climate. It also provides a forum for those who are interested in weather or simply want to keep up with new findings. The society's membership includes meteorologists of standing throughout the world, both amateurs and professionals. Among the society's publications are the *Quarterly Journal of the Royal Meteorological Society*, the *International Journal of Climatology, Meteorological Applications*, and *Weather*.

University Corporation for Atmospheric Research
P.O. Box 3000
Boulder, CO 80307
http://www.ucar.edu/

The UCAR is a consortium of sixty-three member universities with doctoral programs in the atmospheric and related sciences, twenty academic affiliate institutions offering a B.S. and/or M.S., and thirty-eight international affiliates. UCAR manages the National Center for Atmospheric Research (NCAR) and the UCAR Office of Programs. It recently signed a new five-year cooperative agreement with the National Science Foundation, NCAR's primary sponsor. The UCAR mission is threefold: (1) to support, enhance, and extend the capabilities of the university community, nationally and internationally; (2) to understand the behavior of the atmosphere and related systems and the global environment; and (3) to foster the transfer of knowledge and technology to improve the quality of life on our planet.

World Meteorological Organization
7 bis, avenue de la Paix
CH 1211 Geneva 2 Switzerland
http://www.wmo.ch/

The WMO is the authoritative scientific voice on the state and behavior of the earth's atmosphere and climate. From weather prediction to air pollution research, climate change–related activities, ozone layer depletion studies, and tropical storm forecasting, the WMO coordinates global scientific activity to allow increasingly prompt and accurate weather information and other services for public, private, and commercial use. The WMO was created by the World Meteorological Convention, adopted at the Twelfth Conference of Directors of the International Meteorological Organization (IMO) in Washington, D.C., in 1947. Although the convention itself came into force in 1950, WMO commenced operations as the successor to IMO in 1951 and later that year was established as a specialized agency of the United Nations by agreement between the UN and WMO. Its purposes are to facilitate international cooperation in the establishment of networks of stations for making meteorological, hydrological, and other observations and to promote the rapid exchange of meteorological information, the standardization of meteorological observations, and the uniform publication of observations and statistics. The WMO also furthers the application of meteorology to aviation, shipping, water problems, agriculture, and other human activities, promotes operational hydrology, and encourages research and training in meteorology. The WMO is a 185-member organization, with 179 member states and 6 member territories, all of which maintain their own meteorological and hydrological services.

Private Weather Vendor Companies

An updated list containing links to various commercial weather vendors operating in the United States is available from <http://www.nws.noaa.gov/im/more.htm>. The list contains such vendors as Aardvark Weather Software, AccuWeather, Global Atmospherics, WSI Corporation, and Westwynd Consulting. The list does not include individual certified consulting meteorologists.

Chapter Ten

Print and Electronic Resources

This chapter lists books, journals, Internet sites, and other resources related to meteorology and the atmospheric sciences. Information on reference books, general books, and textbooks is followed by overviews of journals and magazines related to meteorological topics. A survey of some useful Internet sites is also presented, along with information on educational videos.

REFERENCE BOOKS

The following resources offer general meteorological reference information and provide excellent supplements to the material presented in the previous chapters.

American Meteorological Society. *Glossary of Meteorology.* **Boston: Author, 2000.**

With over 12,000 definitions, this new edition was produced by an editorial board of 41 distinguished scientists and over 300 contributors. Available in print and CD-ROM.

Burroughs, William J., et al. *Weather.* **Alexandria, VA: Time-Life Books, 1996.**

This easy-to-read and enjoyable text provides remarkable pictures, detailed content, and an extensive resource directory. The book provides an overview of the nature of weather, weather mythology, forecasting, climate change, the relation to human health, and more.

Geer, Ira W., ed. *Glossary of Weather and Climate with Related Oceanic and Hydrologic Terms.* **Boston: American Meteorological Society, 1996.**

A comprehensive glossary of weather and climate terms produced by the American Meteorological Society. Many of the definitions are technical; nevertheless, the book can be a good guide for understanding the jargon of the field.

Kahl, Jonathan D.W. *The National Audubon Society First Field Guide to Weather.* **New York: Scholastic, 1998.**

Written for the general public, the book offers an introduction to weather and weather observing. The book contains a glossary and a resource section, as well as some stunning and educational photos and illustrations.

Longshore, David. *Encyclopedia of Hurricanes, Typhoons, and Cyclones.* **New York: Facts on File, 2000.**

The text is replete with little-known fragments of history, science, and storm chronology. It offers information on individual storms and storm-prone nations and is generally considered a useful and quick reference, despite some scientific and textual inaccuracies.

Miller, E. Willard, and Miller, Ruby M. *Environmental Hazards: Air Pollution—A Reference Book.* **Santa Barbara, CA: ABC-Clio, 1989.**

This useful reference describes and analyzes broad aspects of air pollution, including related environmental hazards. The text contains biographies of people who have contributed significantly to solving air pollution problems and contains a resource listing and a glossary.

Newton, David E. *From Global Warming to Dolly the Sheep: An Encyclopedia of Social Issues in Science and Technology.* **Santa Barbara, CA: ABC-Clio, 1999.**

The book outlines current controversies involving the impacts of science and technology on human life. Topics include global warming and a number of other environmental issues, as well as controversial new developments in the biological, medical, and other sciences. Each entry provides an introductory scientific background, followed by an overview of the science or technology's social, political, and economic implications.

———. *The Ozone Dilemma: A Reference Handbook.* Santa Barbara, CA: ABC-Clio, 1994.
This book offers an unbiased survey of the depletion of stratospheric ozone through overviews of scientific studies, significant papers, and international conferences. The book serves as an excellent foundation for understanding the critical subject of ozone depletion.

Schneider, Stephen H. *Encyclopedia of Climate and Weather.* New York: Oxford University Press, 1996.
This two-volume set contains over 350 scientific articles authored by outstanding contributors in meteorology, physics, chemistry, geology, oceanography, and glaciology. It describes a variety of weather-related processes and contains biographies of meteorologists and other scientists.

Westwell, Ian. *Encyclopedia of the Weather.* Collingdale, PA: DIANE Publishing Co., 1998.
Westwell has compiled a history of meteorology and meteorologists in a format that is both educational and useful. The text is supplemented by high-quality photographs and comprehensive resource information.

GENERAL BOOKS

The following books provide general information on topics in meteorology. Most are written at a level accessible to nonscientists and are either new publications or available in relatively recent editions.

Barnes, J. *Florida's Hurricane History.* Chapel Hill: University of North Carolina Press, 1998.
A reference on hurricanes that have affected Florida, dating as far back as 1546. The book also offers short chapters describing how hurricanes are formed and the damage they can do.

Bluestein, Howard B. *Tornado Alley: Monster Storms of the Great Plains.* New York: Oxford University Press, 1999.
Called "the first serious book on storm chasing," *Tornado Alley* documents Bluestein's experience in pursuing and studying severe storms. The book discusses the development of new and improved

instrumentation and how these tools can help scientists answer how and why some storms produce such deadly tornadoes.

Burroughs, William James. *Does the Weather Really Matter? The Social Implications of Climate Change.* **Cambridge: Cambridge University Press, 1997.**

The book offers an informative account of science and policy for physical scientists as well as for lay readers. The author addresses a number of topics relevant to climate change and societal impacts and includes accounts of historical events and the global warming debate.

Christianson, Gale E. *Greenhouse: The 200-Year Story of Global Warming.* **New York: Walker and Co., 2000.**

Christianson's book offers a rich account of industrialization and its consequences. It covers a wide breadth of topics, from social and political concerns to ideas and controversies surrounding current climate issues.

Davies, Pete. *Inside the Hurricane: Face to Face with Nature's Deadliest Storms.* **New York: Henry Holt and Co., 2000.**

In this look at hurricanes and the people who research them, particularly the scientists at the Hurricane Research Division of the National Oceanic and Atmospheric Administration, Davies examines how hurricane forecasting has advanced and profiles current research methods in a manner that one reviewer likens to "a true tale of high adventure."

de Blij, H.J. *Restless Earth: Disasters of Nature.* **New York: Random House, 1997.**

Written in nontechnical language appropriate for a lay audience, the book highlights many facts and ideas about disturbances of the earth's atmosphere, ocean, and crust. The book reflects the combined writing of many geophysicists as chapter authors and as contributors of individual essays.

England, Gary A. *Weathering the Storm: Tornadoes, Television, and Turmoil.* **Norman: University of Oklahoma Press, 1997.**

Written for lay readers, the book describes some of the major storms in the central United States, offering a good overview of the science and some firsthand reports of tornadoes, their prediction, and resulting devastation.

Faidley, Warren. *Storm Chaser: In Pursuit of Untamed Skies.* Chicago: Independent Publishers Group, 1996.

A clear and understandable text discussing the science of tornadoes and offering autobiographical accounts of storm chasing and tornado encounters. The information is complemented by high-quality photographs from the author's collection.

Fleming, J.R. *Historical Perspectives on Climate Change.* New York: Oxford University Press, 1998.

Fleming tackles the climate change issue from a historical perspective: how we got to the present and what we can learn from the past. The book provides detailed information on the scientific work of five pioneers in the study of the atmosphere and climate and brings readers up to date on global temperature trends in the twentieth century.

Gelbspan, Ross. *The Heat Is On: The Climate Crisis, the Cover-Up, the Prescription.* Boulder, CO: Perseus Books, 1998.

The book investigates the economies and politics of climate change without all the detailed scientific and technical jargon.

Graedel, T.E., and Crutzen, Paul (contributor). *Atmosphere, Climate, and Change.* New York: Scientific American, 1997.

The book contains an overview of how the earth's weather and climate systems work, as well as interesting facts about weather forecasting and climate prediction. It is written in clear, accessible language with many illustrations, diagrams, and tables.

Houghton, John T. *Global Warming: The Complete Briefing.* Cambridge: Cambridge University Press, 1997.

This book is ranked as one of the most comprehensive guides on the science and politics of global warming. It is written in a style accessible to teachers, students, and nontechnical readers alike.

Johnson, Rebecca L. *The Greenhouse Effect: Life on a Warmer Planet.* Minneapolis: Lerner Publishing Group, 1994.

The easy-to-read book, written for students in grades 6 to 9, offers a balanced approach to this topic. The book discusses the basic processes of atmospheric change and the impacts that climate change might have on humans, animals, and plants.

———. *Investigating the Ozone Hole.* Minneapolis: Lerner Publishing Group, 1993.

Johnson's book discusses the science of ozone depletion in the earth's atmosphere, based on observations, interviews with scientists, and firsthand research. The book is geared toward students in grades 5 to 8.

Keen, Richard A. *Skywatch: The Western Weather Guide.* Golden, CO: Fulcrum, 1987.

Full of anecdotes and humor, the book explores the weather of the West in a nontechnical and easily understood style.

———. *Skywatch East: A Weather Guide.* Golden, CO: Fulcrum, 1992.

An easy-to-read book exploring the weather of the eastern United States. The book is written in an easy-to-understand style.

Lamb, H. *Through all the Changing Seasons.* Norfolk, UK: Taverner Publications, 1997.

English meteorologist-climatologist Hubert Lamb (grandson of hydrodynamist Horace Lamb) wrote this autobiography of his education and subsequent career with the British Meteorological Organization. The book contains many anecdotes and provides significant background information on applied meteorology in the mid-twentieth century.

Larson, Erik. *Isaac's Storm: A Man, a Time, and the Deadliest Hurricane in History.* New York: Crown Publishing, 1999.

A blend of science and history that tells the story of the powerful 1900 Galveston hurricane, which killed some 10,000 people and devastated of the city of Galveston, Texas. The story unfolds parallel to the experience of Isaac Cline, at that time a meteorologist with the burgeoning U.S. Weather Bureau.

Laskin, David. *Braving the Elements: The Stormy History of American Weather.* New York: Doubleday, 1997.

Author Laskin merges historical climate information and modern meteorological research into a single, novel-like volume. The book explores the myriad of ways weather has affected us all, from Native Americans, to the first European settlers, to our modern lives.

Monmonier, Mark S. *Air Apparent: How Meteorologists Learned to Map, Predict, and Dramatize the Weather.* Chicago: University of Chicago Press, 1999.

A history of weather maps, from the 1600s to modern satellite imagery. The text is written for general readers but contains significant historical information that may be of interest to professionals.

Parkinson, Claire L. *Earth from Above: Using Color-Coded Satellite Images to Examine the Global Environment.* **Sausalito, CA: University Science Books, 1997.**

The author uses straightforward language to illustrate how satellite images provide information about a variety of geophysical parameters, including snow and ice cover, ozone, sea surface temperature, and vegetation. The book contains fifty-three satellite images as well as maps, photographs, and schematic diagrams and is a useful resource for science classrooms and school libraries.

Pinder, Eric. *Tying Down the Wind: Adventures in the Worst Weather on Earth.* **New York: Putnam Publishing Group, 2000.**

Pinder is a meteorological observer on Mt. Washington, New Hampshire, home to the strongest straight-line winds ever measured on the face of the globe. The book offers an overview of mountain meteorology as well as general information about the earth's atmosphere and climate.

Rosenfeld, Jeffrey P. *Eye of the Storm: Inside the World's Deadliest Hurricanes, Tornadoes, and Blizzards.* **New York: Plenum Press, 1999.**

A detailed look at new developments in storm research written in an entertaining and accessible style. The book offers insights into weather and its causes.

Singer, S. Fred. *Hot Talk, Cold Science: Global Warming's Unfinished Debate.* **Oakland, CA: Independent Institute, 1997.**

The author concisely discusses the climate record; computer models; the effects of clouds, oceans, greenhouse gases, aerosol cooling, and solar variability; and much more. The book is written at a level accessible to laypersons as well as specialists in the field.

Somerville, Richard C.J. *The Forgiving Air: Understanding Environmental Change.* **Berkeley: University of California Press, 1996.**

This well-written and informative text gives an overview of the changes to our atmosphere, as well as the likely effects of some of these changes. By presenting many of the scientific issues in historical context, it helps elucidate how our knowledge continues to evolve.

Sooros, Marvin S. *The Endangered Atmosphere: Preserving a Global Commons.* Columbia: University of South Carolina Press, 1997.

The book is strongly recommended for atmospheric scientists and others interested in the connection between environmental science and public policy. Sooros contrasts current events with those of earlier times and traces the long series of international agreements developed to help preserve our atmosphere and climate.

Stevens, William K. *The Change in the Weather: People, Weather, and Science of Climate.* New York: Delacorte, 1999.

Written by a *New York Times* science reporter, the book examines the evidence for climate change and the apparent increases in severe weather. The accounts are interlaced with a historical perspective and interviews with climatologists.

Verkaik, Arjen, and Verkaik, Jerrine. *Under the Whirlwind.* Toronto: Whirlwind Books, 1997.

The book provides real-life stories of people affected by tornadoes as well as information on dealing with tornadoes before, during, and after they strike. It also contains several high-quality color photographs and sketches, making it an enjoyable book for storm and severe weather enthusiasts.

Wagner, Ronald L., and Adler, Bill (contributor). *The Weather Sourcebook.* 2nd ed. Old Saybrook, CT: Globe Pequot Press, 1997.

A collection of scientific observations and weather folklore as well as almanac-style information on weather and climate patterns. The information is presented in an understandable and friendly fashion, making the text entertaining reading for weather enthusiasts of any level.

Williams, Jack. *The Weather Book.* 2nd ed. New York: Vintage Books, 1997.

Easy-to-understand text and lots of colorful graphics explain how weather happens. The book provides an overview of weather events in the United States, contains climatological information for all states, and discusses the greenhouse effect, ozone hole, and other issues.

TEXTBOOKS

An increasing number of textbooks addressing meteorology at an introductory undergraduate level are available. Some of the most commonly used selections are listed below. A listing of many other print resources is available on-line at <http://www.scd.ucar.edu/dss/faq/print-resources.html>. In addition, each May's issue of *Bulletin of the American Meteorological Society* lists new books, along with new publication announcements and book reviews in each regular monthly issue.

Ahrens, C. Donald. *Essentials of Meteorology: An Invitation to the Atmosphere.* 2nd ed. Minneapolis/St. Paul: West Publishing, 1997.

This book is written for undergraduate students taking an introductory course in meteorology and aims to convey the concepts in a visual, practical, and nonmathematical manner.

———. *Meteorology Today: An Introduction to Weather, Climate, and the Environment/with Infotrack.* 6th ed. Pacific Grove, CA: Brooks/Cole Publishing, 1999.

This text is aimed toward undergraduate students without special prerequisites in science or math. It includes full-color illustrations and photographs to strengthen points and stimulate interest.

Graedel, Thomas E., and Crutzen, Paul (contributor). *Atmospheric Change: An Earth System Perspective.* New York: W.H. Freeman, 1992.

The book explores the physical and chemical workings of the earth's atmosphere at a level accessible for undergraduate students in the sciences. It provides an overview of the science behind current stresses on the earth-atmosphere system.

Lutgens, Frederick K., Tarbuck, Edward J., and Tasa, Dennis (illustrator). *The Atmosphere: An Introduction to Meteorology.* 7th ed. Englewood Cliffs, NJ: Prentice-Hall, 1997.

Designed as a textbook for an introductory undergraduate course, the book is a highly usable tool explaining the basic concepts of meteorology. Each chapter includes a vocabulary list, review questions, and problems highlighting the major concepts.

Nese, Jon M., and Grenci, Lee M. *A World of Weather: Fundamentals of Meteorology*. 2nd ed. Dubuque, IA: Kendall/Hunt Publishing, 1998.

This textbook/laboratory manual covers the basics of meteorology and is available in both regular and three-hole-punched loose-leaf form. Intended readers include nonscience majors taking an introductory course in meteorology.

Turco, Richard P. *Earth Under Siege: From Air Pollution to Global Change*. New York: Oxford University Press, 1997.

The book offers a comprehensive view of atmospheric issues from pollution to ozone depletion to the greenhouse effect. It evolved from Turco's own lectures for an undergraduate-level air pollution course and is written at a level accessible to students of any background.

Wallace, John M., and Hobbs, Peter V. *Atmospheric Science: An Introductory Survey*. San Diego: Academic Press, 1997.

A classic survey text on the atmospheric sciences, written at a level appropriate for undergraduate science majors or beginning graduate students. The book provides an overview of the physical processes of the atmosphere, including atmospheric dynamics and atmospheric radiative transfer.

JOURNALS AND MAGAZINES

American Geophysical Union Journals
2000 Florida Avenue, NW
Washington, DC 20009–1277
http://www.agu.org

Geophysical Research Letters is a semimonthly publication offering short, concise research letters presenting scientific advances likely to have immediate influence on the research of other investigators. Articles relate to a variety of topics in the geosciences.

The Journal of Geophysical Research–Atmospheres (JGR–Atmospheres) publishes articles on the physics and chemistry of the atmosphere, as well as on the atmosphere-biosphere, lithospheric, or hydrospheric interface. Topics pertain to research on earth, the environment, and the solar system. JGR is ranked number 7 of 111 journals in the geosciences.

American Meteorological Society Journals
45 Beacon Street
Boston, MA 02108-3693
http://www.ametsoc.org/

The American Meteorological Society produces several publications discussing issues in meteorology and the atmospheric sciences. Many of these are described briefly below.

Bulletin of the American Meteorological Society is a monthly publication containing papers on historical and scientific topics, including areas of current scientific controversy and debate.

Journal of Applied Meteorology is a monthly publication related to applied research in physical meteorology, cloud physics, hydrology, weather modification, satellite meteorology, boundary layer processes, air pollution meteorology, agricultural and forest meteorology, and applied meteorological numerical models.

Journal of Atmospheric and Oceanic Technology is a monthly publication with articles describing instrumentation and methodology used in atmospheric and oceanic research. Topics include computational techniques, methods for data acquisition, processing, interpretation, and information systems and algorithms.

Journal of Climate is a semimonthly publication containing articles on climate research, including the large-scale variability of the atmosphere, ocean, and land surface; changes in the climate system; and climate simulation and prediction.

Journal of the Atmospheric Sciences is a semimonthly publication related to the physics, dynamics, and chemistry of the atmosphere of the earth and other planets.

Monthly Weather Review is a monthly publication related to analysis and prediction of observed and modeled circulations of the atmosphere, including development of numerical techniques, data assimilation, model validation, and case studies.

Weather and Forecasting is a bimonthly publication containing articles on forecasting and analysis techniques, forecast verification studies, and case studies. The journal reports on changes to the suite of operational numerical models and statistical processing techniques and provides demonstrations of the transfer of research results to the forecasting community.

Royal Meteorological Society Publications
104 Oxford Road
Reading, Berkshire RG1 7LL, UK
http://www.royal-met-soc.org.uk/

Atmospheric Science Letters, a publication that began in June 2000 and is published by the Royal Meteorological Society, contains short articles on issues in atmospheric science.

The International Journal of Climatology is published fifteen times a year with the goal of reporting and stimulating research in climatology. The Royal Meteorological Society and John Wiley copublish the journal, which encompasses global and regional studies of climate, local and microclimatological investigations, changes of climate (past, present and future), and the application of climatological knowledge to a wide range of human activities.

Meteorological Applications is a relatively new journal, first published in 1994. News reports and articles describe applications of meteorological science and encourage information flow between providers and users. The Royal Meteorological Society publishes the journal, which is aimed at applied meteorologists, forecasters, and users of meteorological information and services. The journal is international in outlook, with emphasis on meteorology in European countries.

The Quarterly Journal of the Royal Meteorological Society is acknowledged as one of the world's leading meteorological journals. Results of original research in the atmospheric sciences, applied meteorology, and physical oceanography are published eight times a year. The journal also contains comprehensive review articles, short articles describing minor investigations, and comments on published papers.

Weather is a monthly magazine for all interested in meteorology and is available free to members of the Royal Meteorological Society. The journal contains articles on interesting weather events, a variety of meteorological topics, readers' letters, questions and answers, news from around the world of meteorology, and many photographs. It also offers a Weather Log—miniature daily weather maps covering Europe and the North Atlantic.

OTHER PUBLICATIONS

Consequences
Saginaw Valley State University
7400 Bay Road
University Center, MI 48710

Consequences: The Nature and Implications of Environmental Change is distributed through Saginaw Valley State University and the U.S. Global Change Information Office. The Department of Energy, Environmental Protection Agency, National Aeronautics and Space Administration, National Oceanic and Atmospheric Administration, and National Science Foundation share in funding the journal, which publishes articles related to climate change and climate variability, and their impacts. All issues are available on-line at <http://www.gcrio.org/CONSEQUENCES/introCON.html>.

Earth Interactions
American Geophysical Union
2000 Florida Avenue, NW
Washington, DC 20009–1277
http://www.agu.org/

Earth Interactions is an electronic publication emphasizing research in the earth system sciences. The journal is published jointly by the American Geophysical Union, the Association of American Geographers, and the American Meteorological Society. It contains articles dealing with interactions among the lithosphere, hydrosphere, atmosphere, and biosphere in the context of global issues or global change.

Meteorological and Geoastrophysical Abstracts
550 Newton Road
Littleton, MA 01460

Meteorological and Geophysical Abstracts is distributed quarterly via CD-ROM or as a monthly printed journal. It contains bibliographical information on the world's literature in meteorology, climatology, atmospheric chemistry and physics, and related sciences. For more information, visit <http://www.mganet.org>.

Nature
Nature American, Inc.
345 Park Avenue South, 10th Floor
New York, NY 10010–1707

Nature is a highly respected journal covering all areas of science at a level accessible to people with a moderate background in science. Providing news and research reports, the journal is published in the United Kingdom and offers summaries, new articles, and other information on-line <http://www.nature.com>.

Science
American Association for the Advancement of Science
1200 New York Avenue, NW
Washington, D.C. 20005

Science is considered one of the premier scientific journals for general readers around the world. Each issue contains research articles, news, comments, and other sections. In addition, it offers an extensive web site, including Science NOW, with news stories from the world of science, and Science's Next Wave, a forum discussing topics related to science and to science careers <http://www.aaas.org>.

Science News
231 West Center Street
P.O. Box 1925
Marion, OH 43305
http://www.sciencenews.org/

Science News is considered the best publication available to general readers interested in advances in science. It summarizes important breakthroughs in all fields of science and offers research articles, notes, and book reviews.

Scientific American
415 Madison Avenue
New York, NY 10017–1111

This monthly magazine covering every field of science and technology is the oldest and most prestigious scientific journal written for general readers. It also offers an extensive home page containing weekly features, archives of current and past magazine issues, and question-and-answer sessions between readers and scientists <http://www.sciam.com>. In spring 2000, Scientific American Presents published a special issue on weather.

Weatherwise
Heldref Publications
1319 Eighteenth Street, NW
Washington, DC 20036–1802

Billed as America's only weather magazine, *Weatherwise* is published bimonthly to capture the power, beauty, and excitement of weather. The magazine features vibrant color photographs and well-written, informative articles on weather phenomena, history, and folklore. It is an excellent source for reviews of new books and videos, computer software, and web sites. Weatherwise has its own web site at <http://www.weatherwise.org/>.

INTERNET SITES

The Internet is a wonderful source for a wealth of weather information. Numerous government agencies, universities, private companies, and even individuals maintain excellent reference pages with information on all aspects of meteorology and atmospheric science. Several good documents on global warming and climate issues are also available on the web. A place to start is a report titled "Global Warming Is Here: The Scientific Evidence," by Patrick Mazza and Rhys Roth <http://climatesolutions.org/global_warming_is_here/>.

When on-line, you can find information by typing a keyword or phrase you would like to search for using any search engine. There are usually multiple independent sources addressing most weather and related topics on the web; although a few sources may be unreliable, you can certainly have your choice of reputable sites from which to acquire information on particular topics.

Alistair Fraser's Bad Meteorology page, <http://www.ems.psu.edu/~fraser/Bad-Meteorology.html>. Alistair Fraser of Penn State University offers a site to clear up meteorological myths.
Earthweek, <http://www.slip.net/~earthenv/>. The site, subtitled "A Diary of a Planet," highlights temperature extremes, earthquakes, and other natural events on a weekly basis. Archives are available back to October 1995.
National Library for the Environment, <http://wwwcnie.org/nie/>. The site contains the full text of more than 600 Congressional Research Reports dealing with environmental issues and provides good introductory information for undergraduate and graduate students, faculty, and the general public.
National Oceanic and Atmospheric Administration, <http://www.noaa.gov>. Laboratories affiliated with the NOAA have developed numerous sites providing links and information on a number of topics.
National Severe Storms Laboratory, <http://www.nssl.noaa.gov>. Storm Prediction Center, <http://www.spc.noaa.gov>; and Severe Weather Data page, <http://asp1.sbs.ohio-state.edu/severetext.html>. These sites are excellent for locating information on watches, warnings, and other advisories throughout the United States.
National Weather Service, <http://www.nws.noaa.gov>. The home pages of NWS offices across the country can be particularly useful for finding local weather and climatology information.
Our Planet, <http://www.ourplanet.com/imgversn/>. This site, maintained by the United Nations Environmental Programme, includes material on climate change, ozone, public policy, water resources, and other topics.

Tornado Project, <http://www.tornadoproject.com>. A site to visit for everything you want to know about tornadoes.

UCAR's Weather Page, <http://www.ucar.edu/wx.html>. An excellent place to begin acquiring weather information, with forecasts, weather data, links to other sites, and updates on current weather and atmospheric science topics.

University of Illinois on-line meteorological "textbook," <http://www2010.atmos.uiuc.edu/(Gh)/guides/mtr/home.rxml>. Good introductory and background information on the science of the atmosphere.

Virtual Earth, <http://atlas.es.mq.edu.au/users/pingram/vearth.htm>. Descriptions of web and software resources; a good starting place for general or specific information.

Weather Channel's Storm Encyclopedia, <http://www.weather.com/breaking_weather/encyclopedia>. Well-organized, comprehensive information on all aspects of severe weather.

VIDEOS AND OTHER RESOURCES

Sources like PBS <http://www.pbs.org> have produced various videos on weather phenomena. Many of these are available for purchase from the PBS Store <http://shop.pbs.org>. Titles include the *Raging Planet* series (with individual documentaries on hurricanes, tornadoes, lightning, blizzards, fire, and flood) and *Storm Force*, a set of five videos exploring mighty hurricanes, violent tornadoes, and torrential flooding. NOVA has also produced documentaries on weather and related phenomena, including *Flood!* and *Heaven's Breath: The Power of the Wind*. *Weather Chronicles*, by Chip Taylor (Chip Taylor Communcations), looks at how weather has affected our lives throughout modern history.

A variety of educational kits are also available, including *BoxTopics: Storm and Weather Forecasting*, an educational activity kit for children ages 9 and older. The kit includes materials to build a working weather station; information on storm chasing, weather prediction, and storm formation; an activity book; and supplementary materials. The kit can provide a good introduction to weather and meteorology for children, parents, and teachers.

Other weather resources can be ordered from *The Weather Affects* catalog <http://www.weatheraffects.com>. These include the popular series Tornado Video Classics, and weather instrumentation for the amateur enthusiast.

Chapter Eleven

Glossary

This chapter is partially based on definitions from *Glossary of Weather and Climate with Related Oceanic and Hydrologic Terms* (American Meteorological Society, 1996), edited by Ira Geer. A comprehensive weather glossary compiled by the National Weather Service is available at <http://www.nws.noaa.gov/er/buf/glossary.htm>.

Acid rain—Rain having a pH less than the pH of natural rainwater (5.6). The increased acidity is usually due to the presence of sulfuric acid or nitric acid, often attributed to anthropogenic sources.

Active sensing—The type of remote sensing in which a signal generated by the instrument interacts with the object under observation. Radar and lidar are examples of active sensing techniques.

Advection—Horizontal transport of an atmospheric property (e.g., temperature or moisture) solely by the wind.

Aerosol—A suspension of small particles in the atmosphere. Aerosols include dust, soot, and chemical compounds such as sulfate.

Air quality—A measure of the cleanliness of the air described in terms of the levels of contaminants, especially with regard to their potential effects on human health.

Albedo—The reflectivity of an object, usually expressed as the percentage of radiation returned from a surface compared to the total amount striking it.

Atmospheric pressure—The force per unit area exerted by the weight of a column of air above a given point. The most common pressure unit used on weather maps is the millibar (mb). At sea level, the standard or average value for atmospheric pressure is 1013.25 mb, also expressed as 1 atmosphere (atm).

Baroclinic—A model of the atmosphere in which temperature (and hence density) levels exist on constant pressure surfaces. Baroclinity (or baroclinicity) refers to the layering of air in an atmosphere where surfaces of constant pressure intersect surfaces of constant density. Strong layering or stratification of the atmosphere can give rise to a baroclinic disturbance, or cyclone.

Barotropic—A model of the atmosphere in which density surfaces are parallel to surfaces of constant pressure. Thus, the density and pressure surfaces do not intersect. A barotropic atmosphere was used for the first weather prediction models.

Boundary layer—The layer of the atmosphere in the immediate vicinity of the earth's surface. The boundary layer is usually considered to be the lowest 100 to 3000 meters of the atmosphere.

Chlorofluorocarbon—A synthetic compound containing carbon and halogens, used in the past as foam-blowing agents, aerosol propellants, refrigerants, and solvents because of assumed chemical inertness. CFCs entering the stratosphere are broken down by ultraviolet radiation, releasing chlorine that reacts with and destroys ozone.

Clean Air Act Amendments—Federal regulations setting forth goals for the reduction of sulfur dioxide, nitrogen oxide, and other polluting emissions. These amendments are particularly important to work in air quality and air pollution meteorology.

Climate—The sum of all statistical weather information that helps describe a place or region.

Climate change—A significant change (i.e., a change having important economic, environmental, and social effects) in the climatic state of a locale or large region.

Climate variability—Deviations of climate statistics over a given period of time (e.g., a specific month, season, or year) from the long-term climate statistics relating to the corresponding calendar interval. Climate variability is often used to refer to natural anomalies in the climate record.

Convection—Motions resulting in the transport and mixing of a fluid's properties. In meteorology, *convection* usually refers to atmospheric motions that are predominantly vertical, such as rising air currents due to surface heating.

Convective storm—A storm that owes its vertical development, and possibly its origin, to convection.

Deciview—A unit of visibility defined mathematically in terms of the amount of atmospheric scattering and absorption. In general, a change of 1 deciview can be perceived to be the same on clear and hazy days.

Downdraft—A term applied to the strong downward-flowing air current within a thunderstorm, usually associated with precipitation.

Eddy—Irregular whirls within a fluid, such as air or water. A small volume of air (or other fluid) that behaves differently from the larger flow in which it exists.

El Niño—An irregular or anomalous warming of surface ocean waters in the eastern tropical Pacific. Because the condition often occurs around Christmas, it is named El Niño (Spanish for boy child, referring to the Christ child). In most years, the warming lasts only a few weeks or a month, after which the sea surface temperatures and weather patterns return to normal. However, when El Niño last for many months, more extensive ocean warming occurs and can lead to possible extreme weather conditions in widely separated regions of the globe. *See* ENSO.

Ensemble modeling—A weather forecasting technique in which a numerical model generates several forecasts, each based on a slightly different set of initial conditions. If the forecasts are consistent, they are thought to be reliable. If inconsistent, the forecasts are considered unreliable.

ENSO—El Niño/Southern Oscillation. The term used for the coupled ocean-atmosphere interactions in the tropical Pacific characterized by episodes of irregularly (anomalously) high sea surface temperatures in the equatorial and eastern tropical Pacific. Episodes recur at irregularly spaced intervals (two to seven years), and many persist for as long as two years. *See* El Niño.

Exosphere—The uppermost layer of the earth's outer atmosphere, having a lower boundary at 500 to 1000 kilometers above the earth's surface. It is only from the exosphere that atmospheric gases can, to any appreciable extent, escape to outer space.

Eye—A roughly circular area of relatively light winds and fair weather at the center of a hurricane.

Eye wall—A doughnut-shaped area of intense vertical cloud development and very strong winds that surrounds the eye of a hurricane.

Forecast skill—A statistical score associated with how well a forecast is able to predict a particular variable, usually pressure at a given height. High scores indicate a bad forecast; low scores correspond to a near-perfect prediction.

Front—A boundary or discontinuity separating two different air masses, one warmer and often higher in moisture content than the other.

Fujita-Pearson Scale—A rating system used to designate tornado intensity. Formulated in 1971, this scale uses tornado damage to estimate tornado wind speeds.

Greenhouse effect—The heating effect exerted by the atmosphere on the earth because of the ability of atmospheric gases, primarily water vapor and carbon dioxide, to absorb and emit infrared radiation.

Greenhouse gas—Any gas species that absorbs appreciable terrestrial radiation and contributes to the greenhouse effect in the earth-atmosphere system. The main greenhouse gas is water vapor; others are carbon dioxide, ozone, methane, and nitrous oxide.

Heat island—An area of higher air temperatures in an urban setting compared to the temperatures of the suburban and rural surroundings. It appears as an island in the pattern of isotherms on a surface map.

Hurricane—A tropical cyclonic storm having winds in excess of 115 kilometers per hour. Known in the western Pacific as a *typhoon* and in the Indian Ocean as a *cyclone*.

Initial conditions—Prescriptions of the state of a modeled system at some specified time, usually at the start of the period of interest. The data input to a forecast or climate model from which calculations of the future can be made are referred to as *initial conditions*.

Ionosphere—The region of the atmosphere located between 80 and 400 kilometers above the earth's surface. Derives its name from the ionized gases that reside at these altitudes.

La Niña—An episode of strong trade winds and unusually low sea surface temperatures in the central and eastern tropical Pacific; the opposite of El Niño. La Niña is derived from the Spanish word for girl child. La Niña and El Niño form extremes of the atmospheric trade wind fluctuations called the Southern Oscillation.

Latent heating—A change in temperature due to energy absorbed or released during a change of state (i.e., water vapor to liquid water). An important source of energy in thunderstorms and other atmospheric phenomena.

Lidar (light detection and ranging)—The optical analog of radar. Lidar instruments transmit an intense pulsed light beam (laser) that is reflected by air molecules, tiny particles, and cloud droplets in the atmosphere. The instrument can measure both particle concentration and particle movement (i.e., wind speed).

Linear process—A process that can be represented by a mathematical equation involving no products or roots of variables. In general, linear equations are solvable. Nonlinear equations (such as are used to describe the atmosphere) must often be approximated as linear equations so that a solution can be obtained. *See* Nonlinear process.

Mesoscale—An atmospheric scale that ranges from a few kilometers to about 100 kilometers and includes severe storms and other relatively localized phenomena.

Mesosphere—The layer of the atmosphere above the stratosphere, marked by a rapid decrease in temperature with height.

Meteorology—The study dealing with the phenomena of the atmosphere. Meteorology includes not only the physics, chemistry, and dynamics of the atmosphere, but also many of the direct effects of the atmosphere on the earth's surface, oceans, and life in general. The goals often ascribed to meteorology are the complete understanding, accurate prediction, and artificial control of atmospheric phenomena.

Microburst—An intense downdraft that affects a path typically less than 4 kilometers (2.5 miles) and lasts usually less than 10 minutes. Microbursts may have winds reaching 280 kilometers (174 miles) per hour, and may sometimes be accompanied by precipitation in the vicinity.

Montreal Protocol—A landmark international agreement designed to protect the stratospheric ozone layer by limiting the production and use of ozone-depleting substances (i.e., chlorofluorocarbons, halons, carbon tetrafluoride, and methyl chloroform).

Noctilucent cloud—A rarely seen, wavy thin cloud that occurs in the atmosphere at altitudes between 75 and 90 kilometers (about 45 and 55 miles). Noctilucent clouds resemble cirrus, but are usually bluish white or silvery. They are best seen at high latitudes just before sunrise or just after sunset.

Nonlinear process—A process that can be represented only by a mathematical equation involving products or roots of variables or in which variables occur to powers greater than one or are arguments for trigonometric, logarithmic, or exponential functions. Nonlinear equations are extremely difficult to solve; thus, forecasters must often approximate nonlinear processes as linear in order to predict the state of the atmosphere. *See* Linear process.

Numerical weather prediction—The numerical solution of the fundamental equations of physics applied to forecasting the weather. Numerical weather prediction requires electronic computers and sophisticated computational models.

Operational meteorology—A term used to refer to the production of weather forecasts for real-time decision-making purposes.

Ozone—A molecule of oxygen containing three oxygen atoms. Most of the ozone in earth's atmosphere is located in a layer 20 to 30 kilometers (12 to 19 miles) above the surface. The ozone in this layer is important for blocking harmful ultraviolet radiation from the sun and has been the subject of much research due to a decline in ozone amounts.

Parallel processing—The ability of a computer to carry out multiple tasks simultaneously. Without this ability, weather prediction models would require so much computing time that the forecasts would be outdated by the time they were generated.

Parameterization—The mathematical representation of initial physical data (e.g., temperature, pressure) in a weather or climate prediction model.

Passive sensing—A remote sensing technique that detects energy reflected or emitted by an object being studied without any interference with that object. Many observational satellites rely on passive sensing techniques to obtain useful measurements of the exchange of solar and terrestrial radiation that drives meteorology.

Polar orbiting—A satellite in a sunsynchronous, or near-polar, orbit around the earth. A polar-orbiting satellite will cross the equator at the same local time every day. Polar-orbiting satellites are a source of important observations for weather forecasting.

Pollution—Any alteration of the natural environment by natural or anthropogenic sources creating a condition that is harmful to living organisms.

Radar (radio detection and ranging)—In meteorology, an electronic instrument that broadcasts and receives microwave signals back from targets for the purpose of determining the height, location, movement, and intensity of precipitation areas; based on the property of precipiation particles (rain, snow, hail) to reflect and scatter microwaves.

Radiation budget—The balance of incoming solar and outgoing terrestrial radiation important for driving the weather and climate on earth.

Radiometer—An instrument for measuring solar or terrestrial radiation, used on the ground or aboard aircraft or satellite platforms.

Radiosonde—A small balloon-borne instrument package equipped with a radio transmitter that measures vertical profiles (soundings) of temperature, pressure, and humidity in the atmosphere.

Random forcing—Perturbing or disturbing a weather or climate model by introducing changes in the physical quantities in a random, or unpredictable, fashion.

Remote sensing—The technology of measuring or acquiring data and information about an object or phenomenon by a device that is not in contact with it. Remote sensing techniques, whether from the surface or the ground, are used extensively for measurements of atmospheric properties and quantities.

Resolution—A term related to the spacing at which meteorological variables, used in a numerical model or observed from satellite, apply. A model with 50-kilometer (~30 miles) resolution will provide one temperature estimate valid over an area 50 kilometers by 50 kilometers.

Saffir-Simpson scale—A hurricane intensity scale developed by H. Saffir and R. Simpson to relate possible damage by hurricanes to wind speeds or central atmospheric pressures; 1 is minimal, and 5 is most intense.

Severe thunderstorm—A thunderstorm that produces frequent lightning, locally damaging wind, or hail that is 2 centimeters (0.8 inch) or more in diameter.

Smog—The common term applied to problematic, largely urban, air pollution, with or without a "natural" fog. Degradation of visibility or air quality is almost always implied.

Solar radiation—The total electromagnetic radiation emitted by the sun, with 99.0 percent of the total found in the wavelength band between 0.2 and 4.0 micrometers.

Squall line—A line or narrow band of active thunderstorms.

Stratopause—The boundary between the stratosphere and the mesosphere.

Stratosphere—The zone of the atmosphere above the troposphere, home to earth's ozone layer.

Supercell—Persistent, intense updraft (usually rotating) and downdraft coexisting in a thunderstorm in a quasi-steady state, often producing severe weather, including hail and tornadoes.

Supersonic aircraft—Planes that travel at speeds greater than the speed of sound. Because the planes would travel in the stratosphere where nitrogen oxides in the exhaust could contribute to ozone depletion, widespread development of supersonic aircraft never materialized.

Synoptic scale—The scale used to define weather phenomena over regions from hundreds to thousands of kilometers in areas. This scale lies between the mesoscale and planetary scale.

Teleconnection—A linkage between weather changes occurring in widely separated regions of the globe. Teleconnections are important for long-term weather prediction and for understanding the far-reaching impacts of El Niño and other such phenomena.

Terrestrial radiation—The total long-wave infrared radiation emitted by the earth's surface; sometimes includes the contribution from the atmosphere. Both atmospheric and surface terrestrial radiation play an important role in driving weather and climate processes.

Thermosphere—The region of the atmosphere beyond the mesosphere in which there is a rapid rise in temperature with height. The thermosphere is marked by relatively few atoms and molecules.

Tornado—A column of air rotating rapidly around an almost vertical axis, forming a funnel-shaped or tubular cloud that can come in contact with the ground. Potentially the most destructive of all weather systems, tornadoes have diameters ranging from several meters to over several hundred meters (1 meter = 3.28 feet).

Total column ozone—The number of ozone molecules in an imaginary tube measuring 1 centimeter (0.4 inch) on a side and stretching from the surface to the top of the atmosphere. Total column ozone is measured in terms of Dobson units (DU) where 1 DU = 2.7×10^{16} ozone molecules. The average amount of total column ozone over the earth is about 300 DU.

Tropopause—The boundary between the troposphere and the stratosphere.

Troposphere—The lowermost layer of the atmosphere, marked by considerable turbulence and, in general, a decrease in temperature with increasing height.

Trough—In meteorology, a region of relatively low atmospheric pressure at a given level.

Turbulence—Any irregular or disturbed flow in the atmosphere that produces gusts or eddies.

Upwelling—In the ocean, refers to the rising of water from deep subsurface layers toward the surface. In the atmosphere, the component of radiation (either reflected solar or emitted terrestrial) directed upward from the earth's surface.

Visibility—The greatest distance at which prominent objects can be seen and identified by unaided, normal eyes.

Visual range—The distance at which a black object would just disappear from view.

Vorticity—The tendency of air to rotate in either a cyclonic or anticyclonic manner.

Water vapor—Water in a vapor (gaseous) form. Also called *moisture*.

Weather—The state of the atmosphere at any given time.

Wind profiler—An upward-looking Doppler radar able to measure the vertical profile of horizontal wind speeds.

Index

Note: "i" indicates an illustration; "t" indicates a table.

Acid rain, tropospheric pollution, 41–42
Acid Rain Program, EPA, 42
Active sensing techniques, 77
Advanced Spacebourne Thermal Emission and Reflection Radiometer (ASTER), *Terra* platform, 78
Advanced Weather Interactive Processing System (AWIPS), 65, 66i, 157i
Advection, barotropic model, 6
Aeronomy Laboratory, wind profilers, 68
Aerosol Characterization Experiment in Asia (ACE-ASIA), 86
Aerosols, 83, 134–135; atmospheric radiation research, 135; climate research, 131, 132; cloud physics research, 136
Africa, 96, 115
Aircraft Communications Addressing and Reporting System (ACARS), 10, 70
Air-mass thunderstorms, 14–15
Air-quality, 43; monitoring technology, 139–140

Air pollution: acid rain, 41–43; visibility, 43–44
Air pressure, barotropic model, 6
Air Resources Laboratory: ISIS, 71; SURFRAD, 71
Air temperature: atmospheric regions, 2–4, 3i; barotropic model, 6; global warming, 51–55, 54i
Alabama, thunderstorms, 13–14
American Meteorological Society (AMS): career resources, 163; Certified Consulting Meteorologist, 160, 167; *Curricula in the Atmospheric, Oceanic, Hydrologic, and Related Sciences*, 161; employment opportunity, 167–168, 170, 190i; meteorologist definition, 155; professional society, 236–237; public-private partnership, 197–201; salary information, 168, 170; *2000–2001 Occupational Outlook Handbook*, 168, 170, 192; weather modification policy, 204–207
American Society for the Advancement of Science, 235–236
Angell, Jim, 53

Arctic, SHEBA field experiment, 82–83
Arkansas, tornadoes, 17
Army Signal Corps, weather maps, 5
"Atlantic Conveyor," 24
Atmosphere: composition of, 2, 34, 134–135; regions of, 2–4, 3i
Atmospheric chemistry, 1–2, 33–36, 37i, 38–40; goals of, 137–140
Atmospheric dynamics: definition of, 127; research goals, 127–129
Atmospheric electricity research, 136–137
Atmospheric Environment Program (AEP), 104
Atmospheric gases, measurement of, 73–75
Atmospheric particulate, 34
Atmospheric pressure, 2
Atmospheric radiation, 134–135
Atmospheric science, 1–2, 155
Atmospheric scientists, 159
Australia, El Niño impact, 106
Australian Meteorological and Oceanographic Society, 237
Automated Surface Observing System (ASOS), forecast models, 12, 70
Automobile emissions, air pollution, 45, 46i, 48
Aviation. *See* Commercial aviation
Aviation model (AVN), 8i, 9

Backscatter Ultra Violet spectrometer (SBUV), 74
Balloon-borne Laser in Situ Sensor (BLISS), 74
Balloons, GAINS, 89–91, 90i
Bangladesh, air pollution, 45
Barotropic model, 6
Barron, Eric, 133
Belgium, weather disasters, 97–98
Biosphere, boundary layer research, 134
Blizzards, severe weather, 13, 15, 28
Bluestein, Howard, 20, 141–142
Boundary layer meteorology research, 133–134
Bray, Dennis, 58
Brazil, drought impact, 102

Brennan, Charlie, 57
Britain, weather disasters, 97–98
British Antarctic Survey, ozone depletion, 34–35
British Meteorological Office, global warming, 56
Broadcast meteorologist, 157–158
Brooks, Harold, 17, 104, 142
"Brown Cloud," 45
Bureau of Land Management, air quality monitoring, 44
Bureau of Meteorology, 104

Canada: acid rain, 42; flooding, 100; weather forecasting services, 104
Canadian Meteorological and Oceanographic Society, 237–238
Caracas (Venezuela), weather disasters, 97
Carbon dioxide: food production, 110–111; global warming, 50–51, 51i, 57
Carbon tetrachloride, 38–39
Cavendish, Henry, 34
Certified Consulting Meteorologist, AMS accreditation, 160, 167
Changnon, Stanley, 49, 107, 142–143
Charney, Jule, 6, 62, 143
Chemistry, historical development of, 33–34
Chicago, 45, 115
China: global warming, 117, 132; weather disasters, 96
Chlorofluorocarbons, in ozone layer (CFCs), 34–35, 36, 39
Christy, John, 52, 143
Cicerone, Ralph, 35, 143–144
Cirrus clouds, cloud physics research, 136
Clean Air Act (1977), 44
Clean Air Act Amendments (CAAA) 1990, Title IV, 42
Climate, definition of, 48
Climate history, climate research, 131
Climate models, 55–56
Climate modification, climate research, 131
Climate Monitoring and Diagnostics Laboratory, ozone measurement, 75

Index   271

Climate-observing network, climate research, 131–132
Climate Prediction Center (CPC), 1999 hurricane season, 26
Climate research, advances in, 1–2
Climate variability, 105–109, 129–133
Cline, William, 118
Cloud and Earth's Radiant Energy System (CERES), *Terra* platform, 78, 80
Cloud and Radiation Testbed (CART) sites, 71, 72
Cloud physics research, 135–136
Cloud seeding, 121–122
Clouds: atmospheric electricity research, 137; atmospheric radiation research, 135; boundary layer research, 133, 134; climate research, 131; measurements of, 72–73; role of, 135
Cloudsat, 80
Commercial aviation, 70, 123
Commission on Professionals in Science and Technology (CPST), 168
Computer technology: distributive-shared memory, 63–64; early weather systems, 61–62; impact of, 5, 7, 10, 11
Cooperative Atmosphere-Surface Exchange Study–1999 (CASES–99), 87–88
Cooperative Network for Renewable Resource Measurements (CONFRRM), 71
Crop production, and climate change, 110–114
Crutzen, Paul, 34, 144
Cumulus clouds, thunderstorms, 14
Curricula in the Atmospheric, Oceanic, Hydrologic, and Related Sciences, 161
Czech Republic, acid rain, 42

Daale, Pam, 165–166
Deciviews, air pollution, 43
Decker, Coleen, 163–165
Dengue fever, global warming, 114–115
Denmark, acid rain, 42
Denver, smog, 45
Department of Agriculture: drought monitoring, 28; employment opportunity, 168; UVB Monitoring Network, 72
Department of Defense: ASOS, 68, 70; employment opportunity, 168; NAOS, 89
Department of Energy: CART, 71; employment opportunity, 168
Depolarization and Backscatter Unattended Lidar (DABUL) instrument, 73
Detecting Icing with Polarization Experiment (DIPOLE), 88
Detroit, air pollution impact, 45
Differential absorption lidar (DIAL), 73–74
Dobson, George, 35, 74
Dobson ozone spectrophotometers, 74
Dobson Units (DU), 35
Doppler effect, 66–67
Doswell, Charles, III, 104, 144–145
Douglas, Paul, 5, 145, 157
Downdrafts, atmospheric dynamics research, 128
Drought: consequences of, 102–103; definition of, 101; and epidemics, 115; severe weather, 13, 28, 57, 102; weather disasters, 96
Dynamics meteorology, research goals, 127–129

Earth Observing System (EOS) satellite, 77, 80–81
Earth Science Enterprise, 77, 78, 81
Economic Value of Weather and Climate Forecasts, 105
Eddies, atmospheric dynamics research, 129
Education, in meteorology, 160–162, 172–186, 189, 190i, 191i
El Niño; description of, 29; forecasting of, 30–31, 32–33; impact of, 29, 30–31, 53, 105, 106–107, 108t; 1997–1998 season, 30, 31
El Niño/Southern Oscillation (ENSO), 29–30

Electric discharges, atmospheric electricity research, 137
Electronic Computer Project (ECP), weather forecasting, 62–63
Electronic Numerical Integrator and Computer (ENIAC), 61, 62
Elves, 137
Employment, in meteorological fields, 155–160, 168, 170, 189, 190i, 191i, 191–192
Energy-balance models (EBMs), climate model, 55
Energy budget, atmospheric electricity research, 137
Ensemble modeling, 9–10
Environmental Protection Agency (EPA), 238; Acid Rain Program, 42; air quality monitoring, 44, 47–48; global warming report, 59; ozone measurement, 75; ultraviolet radiation measurement, 72
Environmental Technology Laboratory: Lidar, 67; MMCR, 72. *See also* Wave Propagation Laboratory
Epstein, Paul, 114
Eta model, 8i, 9, 71
Europe, crop productivity, 111, 113
European Centre for Medium-Range Weather Forecasts (ECMWF), 7–8, 8i, 9–10
Exosphere, region of, 4
Eye, of hurricane, 22
Eye disease, UV radiation, 116

Fabian, Peter, 111
Farman, Joseph, 34–35, 36
Farmers, adaptions of, 112–113
Fatalities, weather-related, 193–194, 194t
Federal Aviation Administration (FAA): ASOS, 70; DIPOLE, 88; NAOS, 89
Federal Emergency Management Agency (FEMA): El Niño summit, 30; flood damage, 101
Fennimore, Rob, 81
53rd Weather Reconnaissance Squadron, 25

FIRE III Boundary Layer/Arctic Cloud Component, 88
First ISCCP Regional Experiment (FIRE), 88
Fish and Wildlife Service, air quality monitoring, 44
Flavin, Christopher, 119
Floodplain Management Group, flood risk, 98
Floods: and epidemics, 115; severe weather, 13, 15, 28; weather disasters, 96, 97, 98, 99i, 100–101, 103
Florida, thunderstorms, 13–14
Fog, weather modification, 123
Forecast skill, 11–12
Forecast System Laboratory: ACARS, 70; GAINS, 89; High Performance Computing System, 63, 64i; RUC development, 10; wind profiler, 68
Forrest Service, air quality monitoring, 44
48 Horas Antes de la Tormenta, 167
France, weather disasters, 97
Frost prevention, weather modification, 123
Fujita-Pearson Tornado Intensity Scale, 17, 19, 19t

Gardiner, B. G., 34–35
General circulation models (GCMs), climate model, 56
Germany, 41, 97–98
Glantz, Michael, 106, 145–146
Global Air-ocean IN-situ System (GAINS), 89–91, 90i
Global air temperature, greenhouse effect, 51–55, 54i
Global Change Research Program, climate research, 129–130
Global Climate Change, 55
Global forecast models, 7–8, 8i
Global Ocean Model, 7
Global Ozone Observing System, 74
Global Solutions for Science and Learning, GAINS, 89
Global Spectral Model, 8, 9

Global warming, 50–55, 54i, 56–59, 132, 215–219; agricultural production, 111–112; health issues, 114–117; political responses, 117–120, 120i
Goddard Institute for Space Studies, 56, 162
Gore, Al, 44
Gray, William, 24
Green house gases (GHG): climate research, 131; evidence of, 207–212; and global warming, 50–52, 57, 215–219; and stratosphere, 39
Greenhouse effect, 31, 51
Gulf Coast, thunderstorms, 13–14

Hail storms, severe weather, 15
Hail suppression, weather modification, 123–134
Halon-1211, 39
Hantavirus pulmonary syndrome, 115
"Heat island," 49
Heat waves, impact of, 28, 115–116
High Performance Computing System, NOAA, 63, 64i
High Resolution Doppler Lidar (HRDR), CASES-99, 87
Hong Kong, air pollution, 45
Hood, Robbie, 85
Huff, Floyd, 49
Hulme, Mike, 113
Hurrell, James, 52, 146
Hurricane Andrew (1992), 94, 193
Hurricane Bonnie (1998), 24, 25–26
Hurricane Bret (1999), 26
Hurricane Cindy (1999), 26
Hurricane Earl (1998), 24
Hurricane Floyd (1999), 26
Hurricane George (1998), 24
Hurricane Gert (1999), 26
Hurricane Hugo (1989), 94
"Hurricane Hunters," 25
Hurricane Irene (1999), 26
Hurricane José (1999), 26
Hurricane Lenny (1999), 26–27
Hurricane Mitch (1998), 24, 94
Hurricane season, 23

Hurricanes: atmospheric dynamics research, 127–128; development of, 22; 1998 season, 23–24, 25–26; 1999 season, 26–27; severe weather, 13, 15, 22, 23i

Ice crystals, atmospheric radiation research, 135
Ice storms, severe weather, 13
Illinois-Indiana tornado (1917), 16–17
India: air pollution, 45; global warming, 117, 132; weather disasters, 96
Indian Ocean Experiment (INDOEX), aerosols, 84i, 84–85
Indian Ocean, tropical cyclones, 22, 27
Indonesia, air pollution, 45
Institute for Advanced Studies, ECP project, 61–62
Integrated Surface Irradiance Study (ISIS) sites, radiation measurement, 71
Interagency Monitoring of PROtected Visual Environment (IMPROVE), 44
Intergovernmental Panel on Climate Change, (IPCC); air temperature, 52, 113; global warming, 58, 59, 120; greenhouse gases, 219–227
International Satellite Cloud Climatology Project (ISCCP), 88
Internet sites, 170–171, 259–260
Internship programs, 156, 162
Ionosphere, region of, 4
Iran, air pollution, 45
Irish Meteorological Society, 239
Italy, weather disasters, 97–98

Japan, 22, 78
Jarrell, Jerry, 27
Jefferson, Thomas, 50
Johnson, Bryan, 38, 146

Kalnay, Eugenia, 146–147
Kansas, tornadoes, 17
Karl, Thomas, 115, 147
Keeling, Charles David, 50, 147–148
Keen, Richard, 57
Kelly, John, 168

Kerr, Richard, 53
Kunkel, Kenneth, 108–109
Kyoto Protocol: global warming, 118–119; implementation of, 212–215

La Niña: description of, 24–25, 32; forecasting of, 31, 32–33; and hurricanes, 25; impact of, 32, 105, 108
Langmuir, Irving, 121
Lavoisier, Antoine-Laurent, 34
Lidar, 67
Lightning, atmospheric electricity research, 136–137
Lightning Imaging Sensor (LIS), TRMM, 80
Limited-area Fine Mesh (LFM) model, 6–7
Local forecast models, 7, 8i, 10
Los Angeles, 45
Louisiana, thunderstorms, 13–14, 17

Mahlman, Jerry D., 215
Malaria, global warming, 114–115
Mauna Loa Observatory, carbon dioxide monitoring, 50, 51i
Max Planck Institute for Meteorology, global warming report, 58
McPhaden, Michael, 148
McPherson, Ronald, 9
Measurement technology, boundary layer research, 134
Measurements of Pollution in the Troposphere (MOPITT), *Terra* platform, 78
Medium-range forecast model (MFR), 7, 8i, 10
Melanoma, UV radiation, 116
Meningitis, 115
Menzel, Anne, 111
Mesoscale Alpine Programme (MAP), Lidar, 67
Mesoscale, local forecasts, 10
Mesosphere, region of, 3i, 4
Meteo France, hailstorm forecast, 105
Meteor, term, 1
Meteorological Service of New Zealand, 104, 239

Meteorologist: definition of, 155; employment opportunity, 168, 170; employment resources, 170–171; employment sectors, 155–160, 189, 190i, 191i, 191–192, 192i; education, 160–162, 172–186, 189, 190i, 191i; salary information, 168, 170, 192, 193t
Meteorology: definition of, 1, 160; degree programs, 172–186; degrees received, 189, 190i; internet sites, 259–260; journals/maps, 254–259; professional societies, 235–244; reference/text books, 245–254; videos, 260
Methane gas, greenhouse gases, 50, 51, 57
Methyl chloroform, 38–39
Mexico, air pollution, 45, 46i, 139
Michaels, Patrick, 53
Microbursts, atmospheric dynamics research, 128
Millimeter-wave cloud radar (MMCR), 72
Mission to Planet Earth, NASA, 77
Mississippi, thunderstorms, 13–14
MM5 model, 10
Model Output Statistics (MOS), 7, 9
Moderate Resolution Spectroradiometer (MODIS), *Terra* platform, 78
Molina, Mario, 34, 35, 148
Montreal Protocol on Substances that Deplete the Ozone Layer, 38, 39
Montzka, Stephen, 39, 149
Morales, John Toohey, 166–167
Moritz, Richard, 82
Morson, Berny, 58
Mudslides, weather disasters, 97
Multi-angle Imaging Spectroradiometer, *Terra* platform, 78

National Academy of Sciences, 52, 54, 207–212
National Acid Precipitation Assessment Program, 42
National Aeronautics and Space Administration (NASA), 239–240; CAMEX-3, 85; CFC ozone

Index

depletion, 36; Earth Science Enterprise, 77, 81; employment opportunity, 168; Mission to Planet Earth, 77; NAOS, 89; research career, 159
National Atmospheric Deposition Program/National Trends Network, acid rain, 42
National Center for Atmospheric Research (NCAR), 240; computer system, 63–64, 65; internship program, 162; MM5 model, 10; research career, 158–159
National Centers for Environmental Prediction (NCEP), 8; El Niño forecasting, 30; ensemble modeling, 9; La Niña forecasting, 31; RUC, 10; skill scores, 11i
National Climate Center, climate research, 129–130
National Climatic Data Center, global warming, 53, 56
National Drought Mitigation Center, 28, 102
National Flood Insurance Program, 99, 100
National Hail Research Experiment, 124
National Hurricane Center (NHC), 25, 26
National Meteorological Center, 6, 62. See also National Centers for Environmental Prediction (NCEP)
National Oceanic and Atmospheric Administration (NOAA), 241; Aeronomy Laboratory, 68; Air Resources Laboratory, 71; CAMEX-3, 85; Climate Monitoring and Diagnostics Laboratory, 75; Climate Prediction Center, 26; employment opportunity, 168; Environmental Technology Laboratory, 67, 72; forecast models, 12; Forecast System Laboratory, 10, 63, 64i, 68, 70, 89; global warming, 59; heat threat forecasts, 28; High Performance Computing System, 63, 64i; National Hurricane Center, 25; research careers, 159; satellites of, 81; Wave Propagation Laboratory, 68; weather disasters, 96
National Ozone Expedition (NOZE 1), 36
National Park Service, 44, 72
National parks, air quality/visibility, 44, 47–48
National Renewable Energy Laboratory, CONFRRM, 71
National Resources Defense Council, air pollution impact, 45
National Science Foundation, 89, 241–242
National Severe Storm Laboratory, 13; VORTEX project, 20
National Weather Association, 242
National Weather Service (NWS): annual forecasts, 94–95; ASOS, 70; AWIPS, 65, 66i; barotropic model, 6; career resources, 163; employment opportunity, 168; internship, 156, 162; model initiation, 12; NEXRAD, 66; operational meteorologist, 156; public service, 104
Nested grid model (NGM), 7, 8i, 9
Netherlands, acid rain, 42
New York, air pollution impact, 45
Nitrogen emissions, acid rain, 41–42
Nitrogen oxides, atmospheric electricity research, 137
NOAA Profiler Network, 68
Noctilucent clouds, 57
NOGAPS forecast model, 7
North American Observing System (NAOS), 63, 89
North Central River Forecast Center (NCRFC), Red River flood, 100
Norway, acid rain, 42
Numerical weather prediction models, 4–5, 7

O'Brien, James, 149
Oklahoma City tornado, 19–20
Operational meteorologist, 155–156
Ozone: depletion of, 34–36, 37i, 38, 228–234; measurement of, 74–75,

86; recovery of, 39–40; in troposphere, 40, 45
Ozone hole, 36–37, 37i
Ozone layer, 3i, 3–4
Ozone standard, EPA, 47
Ozonesonde, 74–75, 76i

Pacific Islands, typhoons in, 22
Pacific Northwest National Laboratory, MMCR, 72
Parallel processing, ECP, 62
Parry, Martin, 111–112
Passive sensing techniques, 77
Pennsylvania State University, MM5 model, 10
Peru, El Niño impact, 106
Philadelphia, air pollution impact, 45
Photochemical smog, 45, 46i, 47–48
PICASSO-CENA, 80–81
Pielke, Roger, Jr., 100–101, 119, 149–150, 201
Pielke, Roger, Sr., 58
"Planned and Inadvertent Weather Modification," AMS, 204–207
Pollutants, standards, 47
Prabhakara, Cuddapah, 53, 150
Prairie Skies, 157
Precipitation Radar (PR), TRMM, 78, 80
Pressure, atmospheric, 2
Priestly, Joseph, 34
Private sector, operational meteorologist, 156, 159–160
Private weather vendors, 244
"Public/Private Partnership in the Provision of Weather and Climate Services, The," AMS, 197–201

Radar, 66
Radiation, cloud physics research, 136
Radiation budget, 134, 135
Radiation measurements, 71–72
Radiative-convective models (RCMs), climate model, 55–56
Radiometers, 71
Radiosonde network, origin of, 5
Rapid Update Cycle (RUC), 8i, 10, 71
Rasmussen, Erik, 20

Red River floods (1997), 99i, 100–101
Regional forecast models, 7, 8i, 8–9
Regional Impacts of Climate Change-An Assessment of Vulnerability, The, 59
"Remarks at the Forum on International Geosciences at the National Academy of Sciences," 207–212
Remote sensing, techniques of, 77
Remote sensors, boundary layer research, 134
"Report to the Fifth Conference of the Parties of the United Nations Framework Convention on Climate Change," 212–215
Reversing Acidification in Norway (RAIN), 42
Rosenzweig, Cynthia, 110, 111–112
Rowland, F. Sherwood, 34, 35, 150
Royal Meteorological Society, 242

Saffir-Simpson Hurricane Scale, 22–23, 27
SAGE III Ozone Loss and Validation Experiment (SOLVE), 86
Salary range, meteorological fields, 168, 170, 192, 193t
Santiago (Chile), smog, 45
Satellite Broadcast Network (SBN), 7
Satellite technology, 75, 77–78, 79i, 80–81
Schabel, Matthias, 53
Schaefer, Vincent, 121
Schneider, Stephen, 119, 150
"Scientific Assessment of Ozone Depletion," WMS, 228–234
"Second Assessment Synthesis of Scientific-Technical Information Relevant to Interpreting Article 2 of the UN Framework Convention on Climate Change," 219–227
Severe weather: costs of, 95–96; extreme conditions, 28; hurricanes, 22–28, 23i; thunderstorms, 13–15, 16i; tornadoes, 15, 17, 18i, 19–21
Shanklin, J. D., 34–35
Shuman, Frederick G., 62–63

Singer, Fred, 55
"Six Heretical Notions About Weather Policy," 201–204
Skill scores, 11i, 11–12: computer technology, 62–63
Skin cancer, UV radiation, 116
Smog, 45, 46i, 47–48
Solomon, Susan, 36, 151
Southern Oscillation (ENSO), 29–30, 32–33
Spain, weather disasters, 97–98
Spatial resolution, 8
Spencer, Roy, 52
Sprites, 137
Squall line storms, thunderstorms, 15
Stephens, George, 81
Stolarski, Richard, 35, 151
Storm formation, CAMEX-3, 85
Storm tracks, atmospheric dynamics research, 129
Stratocumulus clouds, cloud physics research, 136
Stratopause, 3i, 4
Stratosphere: climate research, 131; and green house gases, 39; ozone layer, 34–36, 37i, 38–40; region of, 3i, 3–4
Stratospheric Aerosol and Gas Experiment (SAGE), 74, 86
Sulfate aerosol, global warming, 117, 132
Sulfur emissions, 41, 42, 44
Suomi, Verner, 75
Supercell storms, thunderstorms, 15, 16i, 20
Surface Heat Budget of the Arctic Ocean (SHEBA) field experiment, 82–83
Surface Radiation Budget Network (SURFRAD) stations, 71, 73
Sweden, acid rain, 42
Switzerland, weather disasters, 97–98
Synoptic data, 156
Synoptic meteorologist, 156

Taylor, Mike, 57
Teleconnection patterns, atmospheric dynamics research, 129
Television and Infrared Observational Satellite (TIROS I), 75, 77, 81
Tennessee, tornadoes, 17
Terra, EOS, 77–78, 79i
Thermosphere, region of, 3i, 4
Third Convection and Moisture Experiment (CAMEX-3), 85
Third European Stratospheric Experiment on Ozone (THESEO 2000), 86
Third Meeting of the Conference of the Parties to the Framework Convention on Climate Change (COP-3), 118
Thomas, Gary, 57, 151
Thunderstorms, 13–15
Title IV, CAAA, 42
Topography, atmospheric dynamics research, 128
Tornadoes: definition of, 15; Oklahoma City, 19–20; research, 20–21; seasonal frequency of, 16; severe weather, 13, 15, 18i
Total Ozone Mapping Spectrometer (TOMS), 74
Total Sky Imager (TSI), 73
Trace gases, 34, 50
Trade winds, 29–30
Trenberth, Kevin, 52, 152
TRMM Microwave Imager (TMI), 80
Tropein, 2
Tropical cyclones, 22, 27
Tropical diseases, global warming, 114–115
Tropical-extratropical interactions, atmospheric dynamics research, 128
Tropical Ocean Global Atmosphere-Tropical Atmosphere-Ocean (TOGA-TAO) program, 30
Tropical Rainfall Measurement Mission (TRMM), satellite precipitation measurement, 78–79
Tropical storm Arlene (1999), 26
Tropical storm Charley (1998), 24
Tropical storm Emily (1999), 26
Tropical storm Frances (1998), 24
Tropical storm Harvey (1999), 26
Tropical storm Katrina (1999), 26
Tropopause, 2, 3i

Troposphere: pollution of, 40–48; region of, 2–3, 3i, 4
Tropospheric Cloud Component, 88
Tropospheric Ozone Production about the Spring Equinox (TOPSE), 87
Turbulence, boundary layer research, 133
2000–2001 Occupational Outlook Handbook, 168, 170, 192
Typhoons, 22

Ukraine, weather disasters, 96
Ultraviolet (UV) radiation, 71–72, 116
Uncertainties in Climate Change Modeling," Senate testimony, 216–218
United Kingdom Meteorological Office, 7–8, 104
United Nations Economic Commission for Europe, timber industry, 98
United States: air pollution impact, 45; air quality trends, 47–48; atmospheric research budget, 192–193, 193i; drought in, 102, 103; floods in, 98, 101; hurricanes in, 22; severe weather in, 95; tornadoes in, 17; weather-related damage, 193–194, 195t; weather-related industries, 194, 195t
Urbanization, 48–49, 102
UV photometric ozone analyzer, 75
UVB Monitoring Network, 72

Venezuela, weather disasters, 97
Venticinique, Martin, 162–163
Verification of the Origins of Rotation in Tornadoes Experiment (VORTEX) project, 20–21, 21i, 82
Videos, meteorology, 260
Visibility, air pollution, 43
Visible Infrared Scanner (VIRS), TRMM, 80
Volga, weather disasters, 96
von Storch, Hans, 58, 152
Vorticity, 6

Wallace, John, 54–55
Water droplets, cloud physics research, 136

Water vapor: atmospheric radiation research, 135; climate research, 131; measurement, atmospheric dynamics research, 128
Watson, Robert T., 212
Wave Propagation Laboratory, wind profilers, 68. *See also* Environmental Technology Laboratory
Weather disasters, 96–97
Weather forecast models: accuracy of, 11i, 11–12; classes of, 7–10; early, 4–5, 6–7
Weather forecasts: advances in, 1, 126; history of, 5; prediction techniques, 126, 140; process of, 5; seasonal, 128, 130; value of, 94–95, 104–105
"Weather girls," 157
Weather maps, origin of, 5
Weather modification, 121–124
Weather producers, 158
Weather satellites, 5
Weather services, privatization of, 104
Weather Surveillance Radar (WSR-88D), 66–67, 68
Weatherhead, Elizabeth, 39, 152–153
Wentz, Frank, 53
Western Europe, weather disasters, 97–98
Wildfires, 102
Wind Profiler Demonstration Network (WDPN), 68
Wind profilers, 68, 69i
Wind shear, thunderstorms, 14–15
Winter storms, severe weather, 15, 28
Wirth, Timothy E., 207
World Meteorological Organization, 52, 243
World Meteorological Society (WMS), 238–234
Wurman, Joshua, 19–20, 153

Yangtze River flood, weather disasters, 96
Yellowstone Park fire (1988), 94

Zebiak, Stephen, 31
Zeilcik, Mark, 57
Zipser, Edward, 153

About the Author

AMY J. STEVERMER has been involved in numerous scientific writing and editing projects through her work with the Cooperative Institute for Research in Environmental Sciences at the University of Colorado at Boulder and the National Oceanic and Atmospheric Administration. She has also written material for an online science education journal and for other audiences.

GETTYSBURG COLLEGE

3326800 0378015 7

WITHDRAWN

DATE DUE